T0201881

Beyond Chance and Credence

Beyond Chance and Credence

A Theory of Hybrid Probabilities

WAYNE C. MYRVOLD

OXFORD
UNIVERSITY PRESS

OXFORD
UNIVERSITY PRESS

Great Clarendon Street, Oxford, OX2 6DP,
United Kingdom

Oxford University Press is a department of the University of Oxford.
It furthers the University's objective of excellence in research, scholarship,
and education by publishing worldwide. Oxford is a registered trade mark of
Oxford University Press in the UK and in certain other countries

© Wayne C. Myrvold 2021

The moral rights of the author have been asserted

First Edition published in 2021

Impression: 3

Published in the United States of America by Oxford University Press
198 Madison Avenue, New York, NY 10016, United States of America

British Library Cataloguing in Publication Data
Data available

Library of Congress Control Number: 2020950770

ISBN 978–0–19–886509–4

DOI: 10.1093/oso/9780198865094.001.0001

Printed and bound by
CPI Group (UK) Ltd, Croydon, CR0 4YY

Contents

List of Figures and Table

Figures

Table

Acknowledgments

Over the course of writing this book, I have given talks on some of the contents at various places, and I am grateful to have received feedback on those occasions. The book was also read in draft form in a reading group at the University of Western Ontario. I express my thanks for their feedback to members of that group, and to the graduate students in a course for which it was used as a text. Among those who have provided valuable comments on drafts of the book are Dimitrios Athanasiou, Michael Cuffaro, Thomas de Saegher, Lucas Dunlap, Sona Ghosh, Marie Gueguen, Bill Harper, Marc Holman, Haiyu Jiang, Molly Kao, Adam Koberinski, Joshua Luczak, Mackenzie Marcotte, Tim Maudlin, Vishnya Maudlin, John Norton, Filippos Papagiannopoulos, Stathis Psillos, Brian Skyrms, Chris Smeenk, and Martin Zelko. Particular thanks are due to David Wallace and Marshall Abrams for their detailed comments. I thank Zili Dong and Sona Ghosh for assistance with copy-editing and the index.

Part of the writing of this book was done while the author was a visiting scholar at the Pittsburgh Center for the Philosophy of Science. I thank the Center for their hospitality, and for fostering a lively intellectual atmosphere. Research for the book was sponsored, in part, by Graham and Gale Wright, who generously fund the Graham and Gale Wright Distinguished Scholar Award at the University of Western Ontario.

Preface

This is a book about the use of probability in science, with an emphasis on its use in physics, and, particularly, in statistical mechanics. Its impetus stems from a puzzle. The application of probabilistic concepts in physics is ubiquitous (more so than might at first appear, as I argue in the first chapter). This raises the question of how to interpret the probabilities being used. It has long been recognized (and is amply documented by Ian Hacking in his book *The Emergence of Probability*) that the word "probability" has been used in two distinct senses. One of these is an epistemic sense, having to do with degrees of belief of agents, such as ourselves, with less than total knowledge of the world. Following standard usage in philosophy, I will call epistemic probabilities *credences*. In another sense, called by Hacking the *aleatory* sense, probabilities are features of certain sorts of physical set-ups: dice throws, roulette wheels, and the like. We routinely raise questions about whether a coin toss is fair, or whether a roulette wheel is biased, and treat these questions as questions about the objects, questions that can be addressed by experimentation. Again, following standard usage, I will call these probabilities *chances*, or *objective chances*.

It would be a mistake to think of these two senses in which the word "probability" has been used as rival interpretations of probability, one of which must be accepted to the exclusion of the other. They are simply distinct (albeit intertwined) concepts, and each has a legitimate use. Indeed, much confusion could have been prevented if the word "probability" had never been used ambiguously, and "chance" and "credence" used instead.

However, there is a puzzle here. For much of the modern period, it was taken for granted that the fundamental laws of physics are deterministic. Even today, when it is accepted that the world is quantum, not classical, it is far from noncontroversial that the lesson to be drawn from the empirical success of quantum theory is indeterminism at the level of fundamental physics. On the most straightforward conception of what an objective chance would be, such chances are incompatible with determinism. It is this that led writers on probability theory such as Bernoulli (1713), Laplace (1814), and others, to declare that probability is wholly epistemic. And yet these same writers fall, on occasion, into taking it as a matter of physical fact whether a given game of chance is fair, and thus there is a tension in their writings.

One of the main theses of this book is that the familiar dichotomy of *chance* and *credence* is inadequate, and that there is a need for a concept of probability that makes sense in a deterministic context and which combines epistemic and physical considerations. These hybrid probabilities, which are neither wholly epistemic nor wholly physical, I call *epistemic chances*. I argue that they fulfill the role required of probabilities in games of chance and in statistical mechanics.

In order to do so, some ground-clearing is required, as I need to explain why there aren't ready-made serviceable notions that fill the role for objective chance in a deterministic context. One of these, connected with classical probability theory, is the idea that defining the probability of an event is a simple matter: the probability of an event A is just the ratio of the number of possibilities compatible with the event's occurrence to the number of all possibilities. The other is the idea that probability statements can be cashed out as implicit references to relative frequencies in some actual or hypothetical sequence of events. Neither of these attempts at defining probabilities achieve its purpose. These approaches have been extensively criticized, but I do not presume that all of my readers are familiar with these criticisms, and, indeed, expect that some readers will begin their reading of the book as proponents of one or the other of these views. For that reason, I explain the reasons why I think that neither of these approaches provide an adequate conception of probability in Chapter 3, and why my readers should, too.

One of my motivations for writing the book is a dissatisfaction with the way probabilities are treated in much of the philosophical literature. Too much of the literature in the philosophy of probability is lacking in historical perspective, and, as a consequence, debates that were carried out in the nineteenth century are recapitulated in the twenty-first. Where historical reference is made, it is often incorrect. A common narrative has it that Laplace (1814) naïvely thought that probabilities could be straightforwardly defined on the basis of a Principle of Indifference, and that Bertrand (1889) shattered some sort of orthodoxy with his "paradox." The truth is that the examples invoked at the beginning of Bertrand's *Calcul des Probabilités* were meant to illustrate a by-then familiar point made already by Laplace, and that Bertrand was, in fact, echoing Laplace. I hope to restore some historical perspective to contemporary discussions.

The literature on the foundations of statistical mechanics has been heavily influenced by the Ehrenfests' encyclopedia article (1912). That article has considerable merits, but historical accuracy is not one of them. The Ehrenfests sought to secure the place of their late friend and mentor, Ludwig Boltzmann, as the chief architect of statistical mechanics; and Boltzmann's predecessors, in particular James Clerk Maxwell, and Boltzmann's successor, J. Willard Gibbs, who built on and extended Boltzmann's work, receive short shrift. The narrative in the minds of many working in the philosophy of statistical mechanics is that, at the time Boltzmann published his celebrated H-theorem (1872), those working in the field we now call statistical mechanics presumed that a monotonic relaxation towards equilibrium could be derived from the laws of mechanics alone, and that the reversibility objection stemming from Loschmidt (1876) came as a bolt from the blue. In fact, the argument from the reversibility of underlying mechanical laws to the conclusion that irreversibility at the macroscopic level had to be a matter of probability had been familiar to the British physicists, Maxwell, Kelvin, Tait, and Rayleigh for almost a decade before Boltzmann came to this realization. Another effect of the Ehrenfests' influence is the notion that there are two incompatible rival approaches to statistical mechanics: Boltzmannian and Gibbsian. What is now usually counted as Boltzmannian (or neo-Boltzmannian) statistical mechanics consists of extracting one strand from the many to be found in Boltzmann's work. Gibbs saw himself, not as offering a rival approach to Boltzmann's, but as building upon another strand of Boltzmann's work. One of the seminal textbooks of statistical mechanics, that of R. C. Tolman (1938), weaves together considerations drawn from Boltzmann and Gibbs, as do many of its successors. In my presentation, the reader is introduced to both a Gibbsian and neo-Boltzmannian approach; my own view is that the supposed tensions between them have been exaggerated.

My chief motivation for writing the book is the fact that I frequently find, in discussions with my own students and with students that I encounter at conferences and workshops, that a great deal of material that I regard as prerequisite for doing serious, professional-level work on the foundations of probability and on the foundations of thermodynamics and statistical mechanics, is not material that one can presume is familiar to those wishing to embark on such work, and is not readily available. This is a book that I wish had existed, when I was a graduate student and then a junior researcher, beginning to delve into such matters.

The book is meant to be more or less self-contained, and can serve as an introduction to the issues discussed for those not already familiar with them. It can be used (as I have done with an earlier draft) as a text in a graduate course on philosophical issues in probability, for students without background in probability theory. For this reason, I have included, as an Appendix, a brief introduction to probability theory, and include also chapters introducing the basics of thermodynamics and statistical mechanics. These contain material that I think everyone interested in the foundations of these subjects should know, and which is not easy to glean from the existing philosophical literature on thermodynamics and statistical mechanics.

I invite readers to think of this book as a choose-your-own-adventure book. For those who wish to read through the work with a minimum of mathematics, I have relegated most technical details to appendices. But I hope that some readers will take these pages as an invitation to embark on serious work in philosophy of probability. The training of a philosopher, when it comes to relevant background knowledge, is often unsystematic, and we often have to pick up what we need along the way. For graduate students and others without much background in probability, I have included introductory material, and I hope that among my readers will be those who are inclined to work through it.

Note on sources. In this book, I quote many works that were published prior to the twentieth century. The dates and location of first publication are important for historical reasons, and for this reason I have provided references to the original publications. For many of the figures quoted, such as Maxwell, Kelvin, and Boltzmann, there are volumes of collected papers, which makes the works more readily accessible. For convenience, I have provided reference to such collections, in addition to the original publications. For any quotations contained in this book from works written in languages other than English, I have, wherever possible, looked at the original publication, and provide citation of the original for those who want to look at it, as well as the translation from which I am quoting. In cases where an adequate translation was not readily available, the translation is my own, and I have provided the original text in a footnote. I am grateful to Marie Gueguen for assistance with the French translations.

1

The Puzzle of Predictability

1.1 Introduction

It is a familiar fact that some things can be predicted with a reasonable degree
of confidence; others, less so. The daily rising and setting of the sun, the
waxing and waning of the moon, and the motions of the major planets are
familiar examples of the predictable. In other domains, such as the stock
market, the motion of a leaf as it falls from a tree on a blustery day, or the
vagaries of the weather, prediction is harder to come by.

The question I want to ask is: How is prediction possible, in those domains
in which it is?

A venerable answer, and perhaps the first that might suggest itself to some
readers, is: because there are deterministic laws of physics that, together with
initial conditions, uniquely determine the future behaviour of things. This
might or might not be true. But it's irrelevant to the question of what makes
prediction possible.

Even if the underlying laws of physics are deterministic, we are never in
possession of anything even remotely like the sort of specification of the state
of a system that, together with those laws, suffices to determine what we will
observe in the future. We never, in fact, take into account anything more than
a tiny fraction of the physical degrees of freedom of any macroscopic system.
Instead, we engage in radical dimensional reduction of our problems; we
seek relations between a small number of macroscopic variables, discarding
almost everything there is to be known about the system.

It is well known that there are systems that exhibit extremely sensitive
dependence on initial conditions, and there has been extensive literature on
the problems that this poses for prediction. Precise long-term prediction,
for such systems, requires absurdly precise specification of initial conditions.
But the situation we find ourselves in is much worse; we know virtually nothing about the precise state of macroscopic systems to which we successfully
apply our predictive techniques.

Beyond Chance and Credence. Wayne C. Myrvold, Oxford University Press (2021).
© Wayne C. Myrvold.
DOI: 10.1093/oso/9780198865094.003.0001

To get a sense of the magnitude of the problem, consider what might seem to be the easiest case, that of planetary motion. Astrophysicists compute tables of predicted planetary positions with an impressive degree of precision, millennia in advance. The calculation is an intricate one, because it's a gravitational many-body problem, which is tackled by deriving successive approximations to the motion. It is complicated by the fact that, at the level of precision reached by modern astronomy, Newton's law of gravitation is not sufficient, and the theory of general relativity must be taken into account. But never mind all that; suppose you want to get a first approximation to the motions of a single planet, say, Jupiter. The gravitational influence of the sun on Jupiter is by far the strongest factor affecting its motion, and so, a good first approximation can be obtained as a two-body problem.

Suppose then, that you are posed the following problem. You are given the relevant laws of physics, the masses, current positions, and velocities of the Sun and Jupiter, the values of any other macroscopically ascertainable quantities that you might regard as relevant, and permission to ignore everything in the universe other than Jupiter and the Sun, because you are only being asked to provide a first approximation of the future behaviour of the system. You are forbidden to invoke anything else. The only assumption that you may make about the detailed microstate of the system is that it is compatible with the values of the macrovariables that you have been given.

Question: On this basis, what can you say about the future course of the system?
Answer: Virtually nothing.

If you were posed a question like this, you would probably treat it as a gravitational two-body problem. This is exactly solvable, and an approximation to initial conditions yields an approximation to later conditions.

But, if you are treating it as a two-body problem, you are assuming that the Sun and Jupiter will remain, during the interval over which you are interested, roughly spherical bodies small in size compared to the distance between them, and you will treat the problem as one of finding the motion of the centers of mass of these two spheres. You are assuming that they will not fly into pieces. But this is, by the rules stipulated, an illicit assumption. The macrostate of the system is compatible with a wide variety of states on the scale of, say, individual molecules, and, within the scope of this set of states are some that lead to a wide variety of subsequent behaviour. The problem is that it is not a two-body problem; it's at least an N-body problem, where

N is the total number of molecules that collectively make up Jupiter and the Sun.

What we actually know about any macroscopic system doesn't even remotely come close to being enough to tightly constrain the future behaviour of the system. This is what I call the *puzzle of predictability*.

The key to solving the puzzle lies in the phenomenon of *statistical regularities*. It has long been recognized that aggregates of events that are, individually, unpredictable, can give rise to reliable regularities. Thus, for example, the total number of traffic accidents in a given year in Pittsburgh might be more or less constant, from year to year, though exactly when and where the next accident will occur is not predictable. Statistical regularities arise when we aggregate a large number of variables that are effectively random and effectively independent. Counterintuitive as it might seem, effective prediction, even in cases of apparent determinism, *always* involves this sort of statistical regularity. This is why the issue of whether the underlying laws are deterministic is irrelevant to the question of how it is that we can make effective predictions. Even if the laws are deterministic, what is required from them is *effectively random* behaviour of most of the variables that are potentially relevant to a prediction task.

An upshot of this is that considerations that stem from probability theory are required for all—yes, *all*—of our applications of physical theory to prediction, and not only in cases where their presence is apparent because probabilities are explicitly invoked.

Now, a little more detail on these points.

1.2 Determinism and the poverty of our knowledge

In a paper presented to the *Académie Royale des Sciences de Paris* in February 1773, Laplace wrote,

> The present state of the system of Nature is evidently a consequence of what it was at the preceding moment, and if we conceive of an intelligence that, for a given instant, embraced all the relations of the beings of this Universe, it will be able to determine for any given time in the past or the future the respective positions, movements, and generally the properties of all those beings.   (Laplace 1776: §XXV, in *Oeuvres Complètes*, vol. 8, p. 144)[1]

[1] "L'état présent du system de la Nature est évidemment une suite de qu'il était au moment précédent, et, si nous concevons une intelligence qui, pour un instant donné, embrasse tous les rapports des êtres de cet Universe, elle pourra déterminer pour un temps quelconque pris dans

This was, of course, echoed 40 years later in an oft-quoted passage from his *Philosophical Essay on Probabilities* (Laplace 1814: 4; 1902: 4).[2]

Some have attributed to Laplace a vision of the future progress of science in which we humans approach ever closer to the condition of such a being. Nothing could be further from Laplace's intention. He is explaining to his audience why, in spite of the presumed determinism of the laws of nature (taken by Laplace to be an *a priori* truth, based on the principle of sufficient reason), there is need for any such subject as probability theory. The Perfect Predictor is introduced as a *contrast* to the workings of the human mind. Though our predictive successes in astronomy furnish a "feeble idea of this intelligence," the human mind "will always remain infinitely removed from it" (Laplace 1902: 4), and that is why we need probability theory.

This is true for two reasons. The first is that the amount that we actually know is always a minuscule fraction of the complete state of the world. A complete specification of the state would require specification of the state of every atom, every elementary particle. The second is that, even if (*per impossibile*) we could be given this complete state, we could not process the information, could not do the requisite computation. To make any headway, we would have to select a few salient variables, and apply a probabilistic or statistical treatment to the rest.

Our vast ignorance is, of course, only important if what we don't know about is potentially relevant to what we want to predict. One potential solution to the puzzle of predictability that might suggest itself is that, in order to make predictions about certain macrovariables, all we need as input is the values of those macrovariables.

Take the case of planetary prediction. We don't know the details of the internal arrangements of the planets, but, it might be argued, we don't have to. All we need to know about Jupiter is that a certain mass is concentrated, in an approximately spherically symmetric fashion, within a spherical region that

---

le passé ou dans l'avenir la position respective, les mouvements, et généralement les affectations de tous ces êtres."

[2] Laplace is not, in this passage, formulating a novel thesis or saying anything that would have been surprising or shocking to his readers; rather, he is reminding his readers of something that he presumes they take for granted. Indeed, markedly similar passages can be found in earlier writings by Boscovich, Condorcet, and d'Holbach; see Kožnjak (2015) for discussion.

is small compared to interplanetary distances, and that something similar can be said about the Sun and other planets.

This is true; if we know that this holds at the present time, and can be expected to remain true for the foreseeable future, then we have all we need to treat the solar system as an n-body gravitational problem, where n is a relatively small number of astronomical bodies.

However, as already mentioned, what we know about the current macrostate of Jupiter is consistent with its flying apart in short order. To see this, imagine two half-Jupiter sized blobs of gas colliding, and settling down into something like the current state of Jupiter. The relevant physical laws are invariant under velocity reversal, and so, if this is a past that could lead to the observed macrostate of Jupiter, it is also a possible future of the observed macrostate.

It might seem that some fundamental law of physics forbids Jupiter from spontaneously dividing into two half-planets that fly apart from each other. But what? Conservation of energy? Let the resulting half-planets be cooler than Jupiter is now, so that the decrease in internal energy offsets the increase in kinetic energy.

This would involve no violation of the conservation of energy, and need not violate any other conservation law, but it would violate the Second Law of Thermodynamics—at least, as originally conceived (see Chapter 6). This is a key to solving the puzzle of predictability. Understanding the basis for the laws of thermodynamics is requisite for understanding predictability, even in domains where thermodynamics is not explicitly evoked. It was one of the great insights of nineteenth-century physics to realize that the Second Law of Thermodynamics rests on statistical regularities.

Statistical regularities are not the sort of thing that can be subjects of certain knowledge. Unlike regularities underwritten by deterministic laws, they are not exceptionless, and they are at best the sort of thing one can regard as very likely to obtain. The pioneers of the science that came to be called statistical mechanics came to the conclusion that phenomena that would count as violating the Second Law, as originally conceived, should be regarded, not as physically impossible, but only as highly improbable. Thus, what we are seeking is not a criterion for ruling out of consideration the sorts of microstates that would lead to Jupiter-splitting behaviour. What we are seeking is more subtle: some way of regarding it as unlikely that a microstate of that sort will be realized. And this means making sense of probabilistic assertions, even in the context of deterministic physics.

1.3 Statistical regularities and the law of large numbers

Today, statistics is a branch of mathematics, closely connected with probability theory. It was not always so. The original meaning of the word had to do with collection and presentation in a systematic way of demographic data. Here's how the subject was characterized in the inaugural issue of the *Journal of the Statistical Society of London*, published in May of 1838.

> It is within the last few years only that the Science of Statistics has been at all actively pursued in this country; and it may not, even now, be unnecessary to explain to general readers its objects, and to define its province. The word *Statistics* is of German origin, and is derived from the word *Staat*, signifying the same as our English word *State*, or a body of men existing in a social union. Statistics, therefore, may be said, in the words of the Prospectus of this Society, to be the ascertaining and bringing together of those "facts which are calculated to illustrate the condition and prospects of society" and the object of Statistical Science is to consider the results which they produce, with the view to determine those principles upon which the well-being of society depends.
>
> The Science of Statistics differs from Political Economy, because, although it has the same end in view, it does not discuss causes, nor reason upon probable effects; it seeks only to collect, arrange, and compare, that class of facts which alone can form the basis of correct conclusions with respect to social and political government.
>
> (Statistical Society of London 1838: 1)

Nearly 50 years later, in the volume celebrating the jubilee of the society, its president, Rawson W. Rawson, characterized the science as

> the science which treats of the structure of "human society," *i.e.*, of society in all its constituents, however minute, and in all its relations, however complex; embracing alike the highest phenomena of education, crime, and commerce, and the so-called "statistics" of pin-making and London dust bins.   (Rawson 1885: 8)

By that time the mathematical methods that we now think of as the province of the science of statistics had begun to make their way into the profession, though they were not yet dominant. We see this in the Jubilee volume, which contains a paper by F. Y. Edgeworth that introduces his audience to such topics as the normal distribution and significance tests,[3] to the perplexity of some members of the audience (Galton 1885: 266).

Systematic collection of statistics burgeoned in the first half of the nineteenth century, giving rise to (in Hacking's words) an "avalanche of printed numbers" (Hacking 1990: 3) With this avalanche of numbers came an increased awareness of statistical regularities, that is, stable frequencies, in large populations, of the occurrence of individually unpredictable events. These extended, not only to such things as births and deaths, but even to matters such as the occurrences of violent crimes. As Quetelet dramatically put it,[4]

> It is a budget that we pay with frightening regularity, that which we pay to the prisons, labor camps, and scaffolds…every year the numbers have confirmed my predictions, to the point that, I could have said with more exactitude: it is a tribute that man acquits with greater regularity than that which he owes to nature or to the treasury of the State, that he pays to crime!
> (Quetelet 1835: 9)

In England, one conduit of this sort of thinking was Thomas Henry Buckle, described by Theodore Porter as "possibly the most enthusiastic and beyond doubt the most influential popularizer of Quetelet's ideas on statistical regularity" (Porter 1986: 65).

Also in the nineteenth century came an increased awareness that, paradoxical as it might seem, we can see these regularities as arising, not in spite of chance, but *because of* chance. The basic idea is found already in Jacob Bernoulli's *Ars conjectandi* (1713), and has to do with what Poisson (1835; 1837) named the *law of large numbers*.

> Things of any nature are subject to a universal law that can be called the law of large numbers. It consists in the fact that, if we observe very considerable numbers of events of the same nature, depending on constant causes and causes that vary irregularly, sometimes in one direction and sometimes in the other, without their variation being progressive in any given direction,

---

[3] Yes, those existed before Fisher was born!

[4] "Il est un budget qu'on paie avec une régularité effrayante, c'est celui des prisons, des bagnes et des échafauds; …et, chaque année, les nombres sont venus confirmer mes prévisions, à tel point, que j'aurais pu dire, peut-être avec plus d'exactitude: Il est un tribut que l'homme acquitte avec plus de régularité que celui qu'il doit à la nature ou au trésor de l'État, c'est celui qu'il paie au crime!"

we find ratios between these numbers that are very nearly constant. For each kind of things, these ratios have a special value from which they will deviate less and less, as the series of observed events increases further, and which they would reach exactly if it were possible to extend the series to infinity. (Poisson 1837: 7)[5]

We observe stable relative frequencies of events in a certain class, not because there is a Providence watching over the sequence, ensuring that things balance out, but because there isn't. It is the result of aggregation of independent chance events that leads to these stable frequencies.

The basic idea can be illustrated by a coin toss. Suppose a coin is tossed n times, and that the probability of heads on any given toss is p (for a fair coin, $p = 1/2$), and suppose that these tosses are probabilistically independent of each other; the probability of *Heads* on any given toss is p, regardless of the results of the other tosses.[6] Now consider the relative frequency of *Heads*, that is, the fraction, out of all the tosses, of those that land *Heads*. This won't, typically, be exactly equal to p. But for a large number of tosses, it will probably be close to p, and the probability that the relative frequency is more than a given distance from p gets smaller as the number n of tosses increases. See Appendix for further discussion.

1.4 The importation of statistics into physics

This is the intellectual background against which serious work on the kinetic theory of gases began, in the mid-nineteenth century. Statistical regularities were a familiar phenomenon in the social sciences, and there was recognition that the explanation for these regularities was to be based on probabilistic considerations.

---

[5] "Les choses de toute nature sont soumises à une loi universelle qu'on peut appeler la lois des grands nombres. Elle consiste en ce que, si l'on observe des nombres très considérables d'événements d'une même nature, dépendants de causes constantes et de causes qui varient irrégulièrement, tantôt dans un sens, tantôt dans l'autre, c'est-à-dire sans que leur variation soit progressive dans aucun sens déterminé, on trouvera, entre ces nombres, des rapports à très peu près constants. Pour chaque nature de choses, ces rapports auront une valeur spéciale dont ils s'écarteront de moins en moins, à mesure que la série des événements observés augmentera davantage, et qu'ils atteindraient rigoureusement s'il était possible de prolonger cette série à l'infini."

This differs from the corresponding passage in Poisson (1835) only by the insertion of the phrase "de causes constantes" in the second sentence.

[6] A sequence of events of this sort is referred to a sequence of Bernoulli trials.

The kinetic theory of gases posits that a gas consists of a large number of molecules moving in a haphazard fashion. This theory, pioneered by Clausius (though there were important precursors, including Daniel Bernoulli, Herapath, Waterston, Joule, and Krönig), was developed, at the hands of Maxwell, Boltzmann, and Gibbs, into the science that we, following Gibbs' coinage, now call statistical mechanics, and whose scope has been extended well beyond treatment of gases (see Brush 1976 for some of the early history).

Poisson himself, who gave the law of large numbers its name, cites molecular theory as an application. In a body composed of discrete molecules separated by empty space, the intermolecular distances may vary widely. Nonetheless, for sufficiently large volumes, equal volumes will contain roughly equal numbers of molecules (Poisson 1835: 481; 1837: 10). Krönig (1856) also argues for order from aggregated disorder, in accordance with the law of large numbers.

> In comparison with the gas-atoms even the smoothest wall is to be regarded as very uneven, and the path of each gas-atom must therefore be one that is so irregular as to escape calculation. According to the laws of the probability calculus one may assume, however, in place of this complete irregularity, a complete regularity.   (Krönig 1856: 316)[7]

It was James Clerk Maxwell, however, who made the use of statistical methods a central part of kinetic theory. Recognizing that the effect of collisions between molecules would render the details of their motions effectively unpredictable, he adopted the strategy of asking, not for the detailed state of the gas, but for a statistical summary of its properties comparable to the tables compiled by the statisticians.

> I carefully abstain from asking the molecules which enter where they last started from. I only count them and register their mean velocities, avoiding all personal enquiries which would only get me into trouble.
> (Report on a paper by Osborne Reynolds on the flow
> of rarified gases, March 28, 1879, in Harman 2002: 776,
> and Garber et al. 1995: 422)

---

[7] "Den Gasatomen gegenüber ist auch die ebenste Wand als sehr höckerig zu betrachten, und die Bahn jedes Gasatoms muss deshalb eine so unregelmässige seyn, dass sie sich der Berechnung entzicht. Nach den Gesetzen der Wahrscheinlichkeitsrechnung wird man jedoch statt dieser vollkommenen Unregelmässigkeit eine vollkommene Regelmässigkeit annehmen dürfen."

In particular, he asked himself what frequency distribution we should expect for the velocities of molecules of an ideal gas, in conditions of equilibrium—that is, a specification, for any given range of velocities, what proportion of molecules would have their velocities within that range—and the answer he came up with is what we now know as the Maxwell distribution.

In an address to the British Association for the Advancement of Science, Maxwell was explicit that this method was imported into physics from the social science of statistics.

> As long as we have to deal with only two molecules, and have all the data given us, we can calculate the result of their encounter, but when we have to deal with millions of molecules, each of which has millions of encounters in a second, the complexity of the problem seems to shut out all hope of a legitimate solution.

> The modern atomists have therefore adopted a method which is, I believe, new in the department of mathematical physics, though it has long been in use of the section of Statistics. When the working members of Section F [the statistical section of the BAAS] get hold of a report of the Census, or any other document containing the numerical data of Economic and Social Science, they begin by distributing the whole population into groups, according to age, income-tax, education, religious belief, or criminal convictions. The number of individuals is far too great to allow of their tracing the history of each separately, so that, in order to reduce their labour within human limits, they concentrate their attention on a small number of artificial groups. The varying number of individuals in each group, and not the varying state of each individual, is the primary datum from which they work.   (Maxwell 1873b: 440; Niven 1890: 373–74)

This, indeed, is how Maxwell proceeded, when dealing with gases. It was his insight to realize that the interesting question to ask, of a gas, is not its detailed state, but a statistical distribution. When studying diffusion of one gas into each other, one considers, not the detailed trajectories of all the molecules, but matters such as the number of molecules that, on average, are to be found in a given volume large enough to contain an enormous number of molecules, and how this changes with time. Much of his work on the kinetic theory of gases had to do with arguing that the Maxwell distribution is a stable distribution, and, moreover, one that gases would tend to approach if they started out with different velocity distributions.

On this view, the laws concerning the bulk behaviour of gases were to be thought of as statistical regularities, having to do with the aggregate behaviour of large numbers of molecules. The rationale for such a method was the irregularity of the motions of individual molecules, due to the randomizing effect of collisions.[8] This included, for Maxwell, the Second Law of Thermodynamics.

> The truth of the second law is ... a statistical, not a mathematical, truth, for it depends on the fact that the bodies we deal with consist of millions of molecules, and that we never can get hold of single molecule.
>
> > (Maxwell 1878: 279; Niven 1890: 670).

And Maxwell drew the inference that is at the center of this chapter.

> But I think the most important effect of molecular science on our way of thinking will be that it forces on our attention the distinction between two

---

[8] This was of sufficient importance to Maxwell that it warranted being put into verse, to the tune of "Comin' Thro' the Rye" (Campbell and Garnett 1882: 630; 1884: 408).

Rigid Body (sings)

Gin a body meet a body
    Flyin' through the air.
Gin a body hit a body,
    Will it fly? And where?
Ilka impact has its measure,
    Ne'er a ane hae I,
Yet a' the lads they measure me,
    Or, at least, they try!

Gin a body meet a body
    Altogether free,
How they travel afterwards
    We do not always see.
Ilka problem has its method
    By analytics high;
For me, I ken na ane o' them,
    But what the waur am I?

Most of this is comprehensible without translation, but, as the last two lines are crucial to Maxwell's point, here they are:

Every problem has its method
    By analytics high;
For me, I know not one of them,
    But what the worse am I?

That is, there are exact solutions to the problem of finding the trajectories of the bodies, but I am none the worse for not knowing them.

kinds of knowledge, which we may call for convenience the Dynamical and the Statistical.

...

Now, if the molecular theory of the constitution of bodies is true, all of our knowledge of matter is of the statistical kind. ...The smallest portion of a body which we can discern consists of a vast number of such molecules, and all that we can learn about this group of molecules is statistical information. We can determine the motion of the centre of gravity of the group, but not that of any of its members for the time being, and these members themselves are continually passing from one group to another in a manner confessedly beyond our power of tracing them.

Hence those uniformities which we observe in our experiments with quantities of matter containing millions of millions of molecules are uniformities of the same kind as those explained by Laplace and wondered at by Buckle, arising from the slumping together of multitudes of cases, each of which is by no means uniform with the others.

(Maxwell 1873a, in Campbell and Garnett 1882: 439–40; 1884: 361–2; Harman 1995: 819)

1.5 Brownian motion and Langevin equations

To get a sense of how the process of reducing an enormously complicated and intractable problem into a tractable one involving a few key variables and statistical averages over the rest works, consider the case of Brownian motion.

A Brownian particle is moving in a fluid consisting of a huge, macroscopic number of molecules. We can take the fluid to define a background state of rest. Let **V** be the velocity of the particle with respect a reference frame with respect to which the net momentum of the molecules of the fluid is zero. The velocity of the particle will undergo changes due to collisions with the molecules.

Now, if we had to keep track of the detailed motions of all of the molecules, this would be a hopelessly intractable task. What we do, instead, is treat the molecules probabilistically.

Assume that the probability distribution of position and velocities of the fluid molecules is independent of where our particle is. Then, if the particle is at rest with respect to the fluid, it will have no tendency to gain velocity in any direction, as collisions from any direction are equally likely. If, however, it is moving with respect to the fluid, then, on average, the relative velocity of molecules it collides with in the direction of its motion will be greater than the relative velocity of molecules that hit it from behind, and there will be a tendency for it to lose velocity. In addition to this overall tendency to lose any initial velocity it has with respect the fluid, there will also be a rapidly fluctuating force on the particle, with no preferred direction, due to collisions. We can also add in any external forces, due to gravitational or other fields. This gives us the Langevin equation,

$$m\frac{d\mathbf{V}}{dt} = -\gamma\mathbf{V} + \mathbf{F}_{ext} + \mathbf{f} \tag{1.1}$$

Here, γ is a constant, related to the density and temperature of the fluid as well as the size of the particle, $\mathbf{F}_{ext}$ is the net external force, and $\mathbf{f}$ is a fluctuating force term, which is treated probabilistically. As the first two terms represent the mean (average) change in velocity, and $\mathbf{f}$ represents departures from this, the expectation value of $\mathbf{f}$ will be zero (provided, that is, that there is no pressure gradient in the fluid). We assume also that the probability distribution of $\mathbf{f}$ is such that it has finite variance,[9] and that the force at any given time is completely uninformative about the force at any other time. The rationale for treating the force in this way is the assumption that the motions of the molecules of the fluid are so complicated as to be effectively random, and that the velocities of any molecules that collide with our particle are independent of the velocity of the particle.

Omitting the final term yields a deterministic equation. If the external force is zero, its solution is an exponential decay of the velocity, towards a state of rest. When the last term is small compared to the others, we get motion that consists of small fluctuations around solutions to the deterministic equation. But, even in cases where the fluctuations term is not negligible, if we consider a large number of such particles, the mean motion of such particles will closely approximate a solution of the deterministic equation.

This equation can be applied to a visible Brownian particle, with empirical success. An analysis of the same flavor, but more complicated, applies to the motion of individual molecules in a gas, which could, for example, be subject to a gravitational field in addition to their own jostlings. The upshot of an analysis of this sort would be an equation of motion for each molecule involving a term representing the external force and a term that

---

[9] See Appendix if these terms are unfamiliar.

represents the effectively random buffeting by other molecules. We could also apply an analysis of this sort to each molecule in a huge ball of gas, each moving in a gravitational field that is a sum of the gravitational fields of the other molecules and an external gravitational field. The mean motion of all these molecules would give us the motion of the center of mass of the ball of gas.

This is, I claim, what we do when we apply the laws of physics to the motion of Jupiter. The force on each molecule that makes up the huge ball of gas we call Jupiter is a sum of the gravitational forces due to other molecules, the gravitational forces due to external bodies, and a fluctuating force due to collisions with other molecules. The motion of the center of mass is determined by the external forces. If, at some time, the gas is approximately spherically distributed about its center, then the gravitational force due to the other molecules is directed, approximately, at the center of mass of the blob. Each molecule experiences roughly the same external gravitational force. And, if the fluctuating force on each molecule has no net direction with respect to a frame of rest defined by the center of mass of Jupiter, then we can use the law of large numbers to conclude that the ball will remain roughly spherical in shape. We have reduced the problem of dealing with the motion of an incredibly large number of molecules to one involving only the center of mass of the planet.

Things are no different with other applications of deterministic laws to macroscopic objects. Philosophers like to talk about billiard balls colliding with each other. Each billiard ball consists of a huge number of atoms vibrating with thermal motion; interatomic forces tend to keep the body rigid, but this, too, is due to the fluctuations being effectively random and not conspiring to tear the balls apart.

1.6 Why are atoms so small?

In the first chapter of *What Is Life?* (1944), Erwin Schrödinger asked a seemingly nonsensical question: Why are atoms so small? The bare question makes no sense, because atoms are large compared to the Planck length, of size of order unity in atomic dimensions, tiny on the scale with which we usually deal with things, and vastly tinier on astronomical scales. We think of them as small, because they are small on the length-scales that are relevant to our ordinary lives. The meaningful question is: Why are our bodies large compared to atoms?

The reason that Schrödinger gives—and I think he's right about this—is that an organism like us, capable of thought, perception, and purposeful interactions with the world, requires stable regularities for its existence. As he puts it, "what we call thought (1) is itself an orderly thing, and (2) can only be applied to material, i.e. to perceptions or experiences, which have a certain degree of orderliness" (1967: 9). If the organism is composed of atoms and molecules (and what else could it be?), these must be of sufficient number to render its internal workings at least somewhat predictable. And if it is to engage in a purposeful way with the world, it must do so on a level at which there is a reasonable amount of predictability.

Insofar as there are regularities to be found in the world, they are of the statistical sort, involving averages over large numbers of variables that taken individually are effectively random. And this means, to return to our earlier example of planetary motion, even if (*per impossibile*) you were given, not only a complete specification of the physical state of the Sun–Jupiter system at one time, but at a large set of times, to make sense of this you would adopt the procedure of the working members of Section F, and redescribe the surfeit of data in terms of a few salient variables, which, in this case, would include the positions of the centers of mass of the Sun and Jupiter. As with the statistical regularities studies in social sciences, it is in terms of these high-level variables that regularities are to be found.

1.7 The upshot

Applications of physics are shot through and through with probabilistic considerations. In the cases where this is not apparent, because probabilistic considerations are not explicit, this is because the law of large numbers or something like it permits us to disregard statistical fluctuations.

This does not, however, solve the puzzle of predictability, but it allows us to put the question in sharper form, as the answer to the puzzle will require us to answer:

How are we to understand the probabilities that come into play in physics? Are they subjective, objective, or do they partake of aspects of both?

What is the justification for the probabilistic methods that we employ?

In the chapters that follow, I will consider some ready-made answers to these questions, and argue that, though they present us with pieces of the puzzle, none of them provides a fully adequate answer. I will make some suggestions in the direction of a more adequate answer.

1.8 Credits

I first came across the idea that reliable prediction depends on statistical regularities, even in apparently simple physical systems, in Schrödinger's *What Is Life?* (1944). Jenann Ismael (2009) has argued that "a probability measure over the space of physically possible trajectories is an indispensable component of any theory—deterministic or otherwise—that can be used as a basis for prediction or receive confirmation from the evidence" (p. 89). In recent years, David Albert (2015) has emphasized the fact that reliable prediction requires probabilistic considerations. Albert's conclusion is that "An empirically adequate account of a world even remotely like ours in which nothing along the lines of a fundamental probability ever makes an appearance is apparently out of the question" (p. 3). There is something right about this, though I have doubts about the singular—*a* probability distribution, which suggests that there is a unique, correct probability distribution— and also about "fundamental." But I am entirely in agreement that probabilistic considerations are indispensable.

2
Two Senses of "Probability"

2.1 Chance and credence distinguished

As has often been pointed out, the word "probability" is used to cover (at least) two distinct concepts.[1] One concept, the *epistemic* concept, has to do with degrees of belief of a rational agent. The other concept, which Hacking calls the *aleatory* concept, is the concept appropriate to games of chance; this is the sense in which one speaks, for example, of the probability (whether known or not) of rolling at least one pair of sixes, in 24 throws of a pair of fair dice.

Historically, the epistemic use of the word is the older. Before there was a mathematical theory of probability, and before anyone attempted to model the degrees of belief of a rational agent by numerical degrees of belief, one spoke of opinions being more or less probable, in a sense closely related to the etymological roots of the word, which it shares with "approbation." A probable opinion, in this sense, is one worthy of approbation, the sort of opinion that one would find in the works of probable authorities (a locution that sounds odd today). The mathematical theory of what we now call probability has its origins in the treatise of Cardano (written around 1520 but not published until 1663), in the Fermat–Pascal correspondence (1654–60), and in the treatise of Huygens (1657), but none of these authors uses the word "probability" or its cognates in connection with this subject. And rightly so; these works are concerned, not with degrees of belief, but with the fair values of various wagers, conceived as objective facts about what we would call the "chance set-up" involved. It was not until Jacob Bernoulli's *Ars Conjectandi* (published posthumously in 1713) that "probability" (which Bernoulli took in the epistemic sense) became the subject of calculation. Once it did, there was a potential for conflation of the two concepts.

---

[1] For an overview of the history, see Hacking (1975).

Beyond Chance and Credence. Wayne C. Myrvold, Oxford University Press (2021).
© Wayne C. Myrvold.
DOI: 10.1093/oso/9780198865094.003.0002

For that reason, some authors found it necessary, in order to avoid confusion, to point out that the word "probability" has two uses. A particularly clear statement that there are two concepts that need to be distinguished is found in Poisson's book of 1837.

In ordinary language, the words *chance* and probability are nearly synonymous. Most often we will employ one or the other indifferently, but when it will be necessary to distinguish between their senses, we will, in this work, relate the word chance to events in themselves, independently of our knowledge of them, and we will reserve for the word probability the previous [epistemic] definition. Thus, an event will have, by its nature, a greater or less chance, known or unknown; and its probability will be relative to the knowledge we have, in regard to it.

For example, in the game of *heads* and *tails*,[2] the chance of getting *heads*, and that of getting *tails*, results from the constitution of the coin that one tosses; one can regard it as physically impossible that the chance of one be equal to that of the other; nevertheless, if the constitution of the coin being tossed is unknown to us, and if we have not already subjected it to trials, the probability of getting *heads* is, for us, absolutely the same as that of getting *tails*; we have, in effect, no reason to believe more in one than the other of the two events.   (Poisson 1837: 31)[3]

Note that Poisson's use of "chance" refers to single events, and the chance of heads on a coin toss is a matter of the physical constitution of the chance set-up (he says "the constitution of the coin," but clearly it matters also how the coin is tossed). This is not a frequency interpretation.

---

[2] Poisson says "*croix* et *pile*"; *heads* and *tails* is our equivalent.

[3] "Dans le langage ordinaire, les mots *chance* et probabilité sont à peu près synonymes. Le plus souvent nous emploierons indifféremment l'un et l'autre; mais lorsqu'il sera nécessaire de mettre une différence entre leurs acceptions, on rapportera, dans cet ouvrage, le mot chance aux événements en eux-mêmes et indépendamment de la connaissance que nous en avons, et l'on conservera au mot probabilité sa définition précédente. Ainsi, un événement aura, par sa nature, une chance plus ou moins grande, connue ou inconnue; et sa probabilité sera relative à nos connaissances, en ce qui le concerne.

Par exemple, au jeu de *croix* et *pile*, la chance de l'arrivée de *croix* et celle de l'arrivée de *pile*, résultent de la constitution de la pièce que l'on projette ; on peut regarder comme physiquement impossible que l'une de ces chances soit égale à l'autre; cependant, si la constitution du projectile nous est inconnue, et si nous ne l'avons pas déjà soumis à des épreuves, la probabilité de l'arrivée de croix est, pour nous, absolument la même que celle de l'arrivée de pile: nous n'avons, en effet, aucune raison de croire plutôt à l'un qu'à l'autre de ces deux événements. Il n'en est plus de même, quand la pièce a été projetée plusieurs fois: la chance propre à chaque face ne change pas pendant les épreuves; mais, pour quelqu'un qui en connaît le résultat, la probabilité de l'arrivée future de *croix* ou de *pile*, varie avec les nombres de fois ces deux faces se sont déjà présentées."

Cournot (1843) follows Poisson in distinguishing between subjective and objective concepts of probability. As do most of his contemporaries, he avows strict determinism (§39). However, he also introduces a notion of *physical possibility*. Examples of physical impossibility include a heavy cone balanced on its vertex, an impulse applied to a sphere exactly in line with the center, imparting motion without any rotation, and a measuring instrument that measures without any error. Physical possibility comes in degrees.

> In the strict language appropriate to the abstract and absolute truths of mathematics and of metaphysics, a thing is possible or it is not: there are no degrees of possibility or impossibility. But in the domain of facts and phenomenal realities, when two contrary phenomena can be produced and are produced effectively, according to the fortuitous combinations of variable causes with other causes or constant conditions, it is natural to regard a phenomenon as endowed with a greater ability [*habileté*] to be produced, or even all the more possible, in fact or physically, that it occurs in a large number of trials. Mathematical probability becomes the measure of physical possibility, and one of these expressions may be taken for the other.   (Cournot 1843: §44, p. 81)[4]

The word "possibility," Cournot says, is appropriate because it clearly indicates a relation between the things themselves, that is not due to our manner of judging or feeling. Frequency of occurrence in a large number of trials is, for Cournot, an indication of degree of physical possibility, but he is clearly not *identifying* frequency of occurrence with degree of physical possibility.

2.2 Credence-functions and belief states

We turn our attention first to the epistemic conception of probability, or *credence*. The idea behind the epistemic conception of probability is the banal

---

[4] "Dans le langage rigoureux qui convient aux vérités abstraites et absolues des mathématiques et de la métaphysique, une chose est possible ou elle ne l'est pas: il n'y a pas de degrés de possibilité ou d'impossibilité. Mais, dans l'ordre des faits et des réalités phénoménales, lorsque deux phénomènes contraires sont susceptibles de se produire et se produisent effectivement, selon les combinaisons fortuites de certaines causes variables avec d'autres causes ou données constantes, il est naturel de regarder un phénomène comme doué d'une habileté d'autant plus grande à se produire, ou comme étant d'autant plus possible, de fait ou physiquement, qu'il se reproduit plus souvent dans un grand nombre d'épreuves. La probabilité mathématique devient alors la mesure de la possibilité physique, et l'une de ces expressions peut être prise pour l'autre."

observation that there are some things we are fairly certain about, some things about which we are less sure, and others that we strongly doubt. That is, we adopt different epistemic attitudes towards certain things. But should these gradations of belief be represented by numerical *degrees* of belief, and, if they should, why should they satisfy the axioms of probability?

2.2.1   Credence and decision

There is a tradition, stemming largely from the work of Ramsey (1931), de Finetti (1937), and Savage (1954), that takes as its center arch the idea of credences as *guides to action*. One imagines an idealized agent who has preferences between various alternative acts, satisfying certain conditions, and attributes to that agent numerical utilities and credences on the basis of those. To get a flavour of the method, suppose that, for some proposition A, and any numbers N, m, with N larger than m, you had definite preferences between the options,

(a) You receive a cookie if A is true, and nothing otherwise.
(b) You are given m lottery tickets out of a total of N in a lottery (that you regard as fair; you are indifferent between any two tickets). If you win the lottery, you get a cookie, and nothing otherwise.

If you had definite, and coherent, preferences for all of those choices, there would be a real number p such that you preferred the cookie if A whenever m/N is less than p, and preferred the lottery whenever m/N is greater than p.

On this approach, your credence in a proposition A can be thought of as the *betting quotient*, or *betting ratio*, that you take to be fair. This requires a bit of explanation.

Suppose that the worth of various eventualities can be assigned numerical utilities. This means, not only that it makes sense to say that you value one eventuality more than another, but also that locutions such as "the amount by which you value A over B is twice the amount by which you value B over C" make sense.[5] For purposes of illustration, we will, as is commonplace in such discussions, entertain the fiction that utility can be represented by

---

[5] I express it this way because the theory requires only that ratios of utility differences be well-defined, not the absolute value of utilities. That is, the zero-point of utilities is arbitrary, as is the absolute value of any difference in utilities.

money, and that the value, or utility, of a sum of money is constant, that is, independent of how much you already have.[6]

We consider bets in which you lay down a certain sum of money, called the *ante*, for a reward of another sum of money, if some proposition *A* turns out to be true. Suppose, for example, that someone offers you a ticket that entitles you to $100 if it rains tomorrow. How much you would be willing to pay for it is an indication of how confident you are that it will rain tomorrow. If you are certain that it won't rain, you won't pay anything for such a ticket. If you are certain that it will rain, you will regard buying the ticket at any price less than $100 as advantageous to you. If you are less than certain, but nonetheless have in your mind a clear estimate of how likely it is to rain, there will be some threshold price below which you regard the exchange as advantageous to the buyer of the ticket, and above which you regard it as advantageous to the seller. That threshold is what you regard as the *fair price* of the ticket.

In general, if a bettor pays a quantity *R* for a ticket that pays *S* if *p* is true, we will call the ratio *R/S* the *betting ratio*, or *betting quotient*. On the fiction of constant utility of money, the betting ratio you regard as fair is independent of the magnitude of *S*, and we can take your fair betting ratio as a measure of your degree of belief in *p*, or your *credence* in *p*.

A set of judgments about fair betting ratios is called *coherent* if they satisfy the axioms of probability. A case for coherence as a condition of rationality can be made on the principle that two options that pay exactly the same in all circumstances should be valued equally. As Ramsey put it, "If anyone's mental condition violated these laws, his choice would depend on the precise form in which the options were offered him, which would be absurd" (Ramsey 1931: 182). Consider, for example, the axiom of additivity, which says that, for any incompatible propositions *p*, *q*,

$$Pr(p \vee q) = Pr(p) + Pr(q).$$

Let *A* and *B* be incompatible propositions, and consider the two options:

---

[6] It's not, of course. $100 at the end of the month means a lot more to you if it makes a difference between your rent cheque clearing or not, than it would if it's merely the difference between one luxury item and another. But, if we consider only sums that are small compared to what you already have, the fiction of invariant utility of money might be a good enough approximation to serve for illustrative purposes.

(a) A ticket that entitles you to a cookie if A, and a ticket that entitles you to the cookie if B.

(b) A ticket that entitles you to a cookie if $A \lor B$.

Since A and B are incompatible, on either option you get at most one cookie, and you get a cookie if either A or B is true, and not otherwise. Thus, each of these options amounts to the same thing. What you regard as a fair price for option (a) is, presumably, the sum of what you regard as fair prices for each of the two tickets in (a). If, therefore, your judgments regarding fair betting ratios for these options violate the Additivity condition, you will value (a) and (b) differently.

As a symptom of this, you might be willing to buy one option and sell the other at different prices, resulting in a situation in which you have a net loss no matter what happens, which is known as a "dutch book."[7] The dutch book argument is sometimes, misleadingly, presented as an argument of prudential rationality, with the claim being that an agent whose judgments about fair betting ratios violated it would inevitably accept a sure-loss bet if offered it. But, of course, when presented with such an offer, the agent might realize that it is a sure-loss bet, which might give her pause and provide incentive to rethink her judgments about fair betting ratios. Such an agent is acknowledging a flaw in her judgments, which is really the point. The possibility of a dutch book is a symptom that equivalent bets are being valued differently, which is really what is problematic about credences that violate coherence.

---

[7] The argument, mentioned in passing by Ramsey, is spelled out by de Finetti (1937). Neither of them uses the term "dutch book." According to the website *Earliest Known Uses of Some of the Words of Mathematics* (Miller, accessed Aug. 30, 2019),

> The term entered circulation in the 1950s. R. Sherman Lehman explains the term, "If a bettor is quite foolish in the choice of the rates at which he will bet, an opponent can win money from him no matter what happens. The phenomenon is well known to professional bettors—especially bookmakers …Such a losing book is known by them as a 'dutch book.' Our investigations are thus concerned with necessary and sufficient conditions that a book not be 'dutch.' "
>
> ("On Confirmation and Rational Betting,"
> *Journal of Symbolic Logic* 20 (1955): 251)

According to Farmer and Henley's *Slang and Its Analogues Past and Present* (1891), the use in English of "dutch" as an epithet of inferiority is "[a] witness, no doubt, to the long-standing hatred engendered by the bitter fight for the supremacy of the seas between England and Holland in the seventeenth century" (Farmer and Henley 1891: 347). Usage of this sort lingers on in "dutch treat" and "dutch courage."

Farmer and Henley define *dutch bargain* as "a bargain all on one side" (p. 348). It seems likely that this usage is related to "dutch book," as, in a dutch book, the advantage is all on one side.

There is also an accuracy-based argument for the conclusion that an agent's credences should satisfy the axioms of probability. An argument of this type was first formulated by Joyce (1998); see Joyce (2009) for a reformulation in response to criticisms and Pettigrew (2016b: ch. 7) for an alternative argument. The idea is that it is a virtue to have degrees of belief that accord high credence to true propositions, low credence to false ones. An accuracy score is assigned to a set of credences, from a set of scoring rules that satisfy certain criteria of admissibility. Given a partition $\{P_1, \ldots, P_n\}$, a credence function that assigns numerical credences to all of the elements of this partition can be represented as a point in n-dimensional space. One way to assess the accuracy of a credence function is to take the negative of the square of the Euclidean distance between this point and the point that represents the truth, that is, the function that assigns the value one to the true member of the partition, and zero to the rest (we take the negative because we want a higher score to be better). This accuracy score is known as the *Brier score* (Brier 1950), and has been used to rate the accuracy of weather forecasts. This rule is in the family of rules called *proper scoring rules*, meaning that, if you judge credence functions other than your own by their *expected* score, that is, a weighted average of the score under all possibilities, with the weights given by your own credences, then you will judge no credence function to be better than your own. The accuracy-based argument for probabilism proceeds by showing that, for any credence function that violates the axioms of probability, there is another that has a higher accuracy score in any eventuality—that is, no matter what the world is like.

Representation of an agent's credal state by a credence function is often known as *Bayesianism*.[8] Don't be fooled by the term into thinking that it denotes a unique philosophical position. In an attempt to dispel the impression that it did, in 1971 I. J. Good published a letter to the editor of *The American Statistician*, outlining a classification of Bayesian positions according to 11 criteria, which yielded 46,656 varieties of Bayesians (Good 1971). Good's letter did not have the desired effect, and people still talk of "Bayesianism" as if there were only one such thing.

For our purposes, it is worthwhile to classify views according to the breadth of probability assignments they countenance. At one extreme is

[8] The term "Bayesian" seems to have arisen in 1950 or shortly before. As is pointed out by Miller, R. A. Fisher used it in the notes he wrote for his collection of papers, *Contributions to Mathematical Statistics* (1950). Hurwicz used it in a lecture delivered on Dec. 29, 1950, at the meeting of the Econometric Society in Chicago (Hurwicz 1951).

radical subjectivism, which maintains that the only constraint on reasonable credences is that they be *coherent*, that is, that they satisfy the axioms of probability. At the other is *Objective Bayesianism*, according to which there is some principle of rationality that for any body of evidence uniquely determines a single rational credence-function conditional on that body of evidence. In between is are positions of the sort that Shimony (1971) has dubbed *tempered personalism*. On an approach such as this, the class of credences deemed appropriate for a reasonable agent who wishes to learn about the world is narrower than the class of all coherent credences, but no attempt is made to narrow this class to a singleton.

The attitude adopted in this book is a form of tempered personalism. I am unconvinced by attempts to argue for some principle that prescribes unique credences, for reasons that will be discussed in the next chapter. On the other hand, a radical subjectivism that declares all coherent credences to be on a par, and regards differences between them as akin to matters of taste or feeling, goes too far in the other direction. One coherent credence assignment might be better than another, in two ways. For one thing, better-informed credences are better than poorly informed ones. This sort of valuation of informed credences fits well within the subjectivist framework, as it can be proven that, given a choice between making a decision based on your current credences, and gaining information and then making the choice, you should prefer the latter, provided that the expected cost of obtaining the new information does not outweigh the expected increase in the value of your credence function.[9] Also, a credence function ought not to be too dogmatic. A credence function that assigns all (or most) weight to one out of a number of alternatives, in the absence of strong evidence in favour of that alternative, may rightly be judged unreasonable, though it may pass the test of coherence. It is avoidance of excessive dogmatism that motivates Shimony's "tempering condition."[10]

For example: suppose that you have a friend who, while gambling in a casino, becomes completely convinced that on the next spin of the roulette wheel the ball will land on number 23, so completely convinced that he is

---

[9] This stems from a theorem that is originally due to Good (1967), and which has been generalized in various ways; see Myrvold (2012b) for discussion.

[10] The original meaning of the verb *to temper* is to bring something to a desired condition, state, or quality, by mingling it with something else. Its chief connotation these days is moderation, avoidance of extremes. But a metal may be made stronger by alloying it with another substance, as in *tempered steel*. The tempering invoked by Shimony is meant to carry both connotations, of moderation and of making stronger.

about to bet everything he owns on that proposition.[11] Not because he thinks that the wheel is rigged; he believes it to be a perfectly ordinary roulette wheel. Rather, he is convinced, with absolute certainty, that, among the possible physical microstates of the wheel and the croupier and everything else that might influence the outcome of the spin, the actual microstate is in the set that leads to 23. You would not accept his credences as reasonable, and, if you are indeed his friend, you would try to dissuade him from placing the bet. The correct response to such a person would be to say, simply, that no, you don't have that level of detailed knowledge about the microstate; you are being unreasonable. One can form such a judgment without being committed to the idea that there is a *unique* credence that *is* reasonable.

I assume that the reader shares this judgment. For our purposes, it will not be necessary to precisely delimit the class of reasonable credences; judgments that are somewhat vague will suffice. Despite avowals that can sometimes be found in the literature (see e.g. de Finetti 1937: 8; 1980: 64), I do not believe that anyone sincerely judges all coherent credence functions to be on a par. I will not argue here that not every set of credences satisfying the axioms of probability is reasonable for an agent with limited access to information about the physical world. If I ever meet someone who sincerely doubts this, I may try to formulate an argument, but I am not interested in addressing doubts frivolously feigned for the sake of playing devil's advocate.

2.2.2 Conditional bets and conditional credence

One can consider bets that take place only if some condition q is satisfied. Suppose, for example, that you have the opportunity of betting on a proposition p, with the bet called off, and all monies returned to the bettors, if condition q fails to obtain. What is the fair price of such a bet? It might differ from the fair price of an unconditional bet on p, if you take the occurrence or non-occurrence of q to be relevant to p. If, for example, you regard occurrence of p as more probable if q occurs, you should be willing to pay more for a bet on p, conditional on q, than for an unconditional bet, because, in the event that q doesn't occur, you lose nothing on the conditional bet.

---

[11] This example is inspired by James Garner's character Latigo Smith in the 1971 comic western *Support Your Local Gunfighter*.

If you have judgments about fair prices of unconditional bets, these place constraints on your judgments about fair prices of conditional bets, on Ramsey's principle that your judgment about the worth of an option should not depend on which of two equivalent descriptions is used, as any conditional bet is equivalent to a pair of unconditional bets.

Suppose that your fair betting ratio for $p\&q$ is α, and your fair betting ratio for q is β, and hence your fair betting ratio for $\sim q$ is $1 - \beta$. Consider two bets, on $p\&q$ and $\sim q$, respectively, both of which you regard as fair.

(a) Bet αR for a return of R if $p\&q$ is true.
(b) Bet $(1 - \beta)S$ for a return of S if q is false.

Your total ante is $\alpha R + (1 - \beta)S$. If, now, we choose the stakes R and S so that $\alpha R = \beta S$, then your total ante is just S, which is the amount that you win if q is false. If q is false, your win therefore amounts to simply returning your ante to you. That means that the pair of bets is equivalent to the following conditional bet.

(c) Bet S for a return of R if p is true, conditional on q being true. If q is false, the bet is called off, and your ante is returned.

If you regard the pair of unconditional bets (a) and (b) as fair, you should regard the conditional bet (c) as fair also. Clearly, given any conditional bet, we could construct, along the lines of (a) and (b), a pair of unconditional bets that taken together are equivalent to the conditional bet. The betting ratio in (c) is S/R, which, if β is non-zero, is equal to α/β. Thus, if your credence in q is greater than zero, your conditional credence $Cr(p|q)$, which is the fair price of a bet on p conditional on q, satisfies

$$Cr(p \mid q) = \frac{Cr(p\&q)}{Cr(q)}. \tag{2.1}$$

2.2.3 Updating credences

Suppose that you are about to learn whether or not a proposition q is true, and adjust your credences in other propositions in light of this information. On the basis of your current credences, what would you recommend to your future self?

Here's how to think about it. Suppose you want to recommend to your future self a betting quotient on p that would be appropriate to use upon learning q. Provided that you are about to learn whether or not q is true, a bet that your future self makes if and only if she learns that q is true is equivalent to a conditional bet made now, conditional on q. Therefore, what you will recommend to your future self as the credence to adopt upon learning q is your current conditional credence, $Cr(p|q)$.

That is, you will recommend a shift in credence, upon learning q,

$$Cr(p) \quad \Rightarrow \quad Cr(p|q). \tag{2.2}$$

Such a shift is called *updating by conditionalization*, or *Bayesian conditionalization*.

When will such a shift be appropriate? Well, if, in the interim between your initial assessment and the time that you learn q, you have had occasion to rethink your judgments, then you might not continue to endorse your earlier recommendation to your future self. Suppose, however, the shift involves nothing other than learning that q is true, and no reconsideration of prior judgments—a shift that has been called a *pure learning experience*. In such a situation you continue to endorse your earlier judgments, and updating by conditionalization is appropriate.

Provided that one's prior credence in q is non-zero, we have

$$Cr(p|q) = \frac{Cr(q|p)Cr(p)}{Cr(q)}. \tag{2.3}$$

Equation (2.3) is called *Bayes' Theorem*, in accordance with Stigler's Law of Eponymy.[12] It is often employed when considering the impact of evidence e on a hypothesis h, in which case it becomes

$$Cr(h|e) = \frac{Cr(e|h)Cr(h)}{Cr(e)}. \tag{2.4}$$

2.2.4 Imprecise credences: Good's hierarchy

People don't, of course, have arrays of definite preferences rich enough to determine precise numerical credences, and, more importantly, to do so would be absurd. As I. J. Good puts it, "it would only be a joke if you were to say that the probability of rain tomorrow (however sharply defined) is

---

[12] Stigler's Law of Eponymy states that "No scientific discovery is named after its original discoverer" (Stigler 1999: 277). Stigler attributes the Law to Robert K. Merton.

0.3057876289" (Good 1979, in Good 1983: 95). On the other hand, it is quite reasonable for a meteorologist, given sufficient evidence, to form a judgment that the probability of rain is between 0.30 and 0.40.

In addition to preferring one act to another, or being indifferent between them, one might reasonably also be *undecided*, which is not the same as being indifferent. To be indifferent is to regard the two acts of being of equal value; to be undecided is to consider that one doesn't have an adequate basis for making such a judgment. Here's one way to see the difference. Indifference between acts, construed as a judgment that they are of equal value, should be transitive. Indecision need not be. Suppose I have a definite preference between two options, A and B. Someone introduces a new option, C, something unfamiliar, something I have never thought of before. I might regard myself as being in no position to decide between A and C, and *also* in no position to decide between B and C. It does not follow from this that I have become undecided between A and B. Indecision between two options is not a judgment of equal value; it is an absence of judgment, a shrug of the shoulders.

There is a temptation—and a long tradition of succumbing to this temptation, which we will discuss in the next chapter—to think that in a state of complete ignorance concerning which of two alternatives is true, the mere fact of your ignorance requires that you judge them equiprobable. On this reasoning, if I know of no relevant difference between the two sides of a coin, I should judge the event of the coin landing *Heads* when tossed to be equiprobable with the event of it landing *Tails*.

But this is too quick. A judgment of equiprobability is not an absence of judgment. The reasoning tacitly presupposes that there are only three alternatives: to judge *Heads* more probable than *Tails*, to judge *Tails* more probable than *Heads*, or to judge them equiprobable. There is, however, a fourth alternative, the one appropriate to conditions of ignorance, and that is the shrug of the shoulders, indicating that you have no grounds for any of the three judgments. That is, you might be *undecided* between the two propositions. Like indecision in preference, indecision in probability is distinguished from a judgment of equiprobability in that indecision need not be transitive.

One way to represent an epistemic state that has room for indecision is to model an agent's epistemic state via a set of credence functions, which can be thought of as the set of credence functions that the agent regards as reasonable, given the evidence that she has. On this approach, an agent regards a proposition as more likely than another if and only if all credence functions in her credal set agree on the ordering; indecision is represented by having the credal set include functions that disagree on the ordering. A representation theorem, giving conditions under which an agent's credal state can be represented by such a set, has been provided by Seidenfeld et al. (1995).[13]

There is a sense in which this replaces one false precision with another. You might be decided regarding certain options, undecided regarding others, but the cutoff is not a sharp one. There might be some credence functions that you regard as reasonable, others that are unreasonable, but it would be a joke to say that 0.3057876289 is in your credal set for rain tomorrow, but 0.305787629 is not. Instead of a sharp set of credence functions, one might consider reasonableness a matter of degree, and replace the credal set by a second-order probability function, assigning degrees of reasonableness to any candidate credal function.

Of course, this move also replaces one false precision with another; in that it assigns a *precise* degree of reasonableness to any credal function. We might instead consider a *set* of second-order probability functions assigning degrees of reasonableness to first-order credal functions, or, to fuzzify that, a third-order, and so on, as recommended by I. J. Good (1952; 1979). We should not expect the regress to be terminated at some level with a probability function that is entirely determinate; rather, one would expect that, the higher one goes up the hierarchy, the harder it will be to uniquely specify a probability function.

There's a potential infinite regress here. This might make things seem hopeless. It need not, though; as Good remarked in a discussion of decision making with a hierarchical representation of belief states,

> the higher the type the woollier the probabilities. It will be found, however, that the higher the type the less the woolliness matters, provided that the complications do not become too complicated.
>
> (Good 1979, in Good 1983: 14)

Provided that this is right, that the higher the type the less the woolliness matters, one can truncate the hierarchy at a point at which any added benefit of going further up outweighs the cost. The regress need not be a vicious one. Good, once again:

---

[13] For more on representation of credal states via sets of credence functions, see Bradley (2019) and references therein.

Isaac Levi... says, "Good is prepared to define second order probability distributions... and third order probability distributions over these, etc. until he gets tired." This was funny, but it would be more accurate to say that I stop when the guessed expected utility of going further becomes negative if the cost is taken into account.   (Good 1979, in Good 1983: 99)

It is sometimes useful to represent a person's credal state via numerical credences. Normative considerations lead to the conclusion that violations of the axioms are departures from ideal rationality, which we might disregard if we are engaged in normative epistemology. For other purposes we should take indecision seriously, and it can be useful to represent an agent's credal state via a set of credence functions. We do not thereby depart from the realm of normative epistemology. For still other purposes, it may be useful to proceed to the next step of Good's hierarchy. Think of all of these representations as formal tools, useful for certain purposes. No claim is being made about psychological reality of these representations of credal states.

A representation of a state of belief via a set of credence-functions furnishes a built-in distinction between indecision and a judgment of equiprobability. If you judge two propositions p and q to be equiprobable, then we will represent this by having every credence-function in your credal set accord them equal probability. However, if you are undecided about the two propositions, this is represented by a credal set that contains both credence-functions that favour p over q, and credence-functions that favour q over p.

There is a tradition within philosophy of trying to imagine what the credences would be of a fictional creature that somehow was in command of a language rich enough to express the propositions of interest, versed in logic and the probability calculus, but in a state of complete and total ignorance about the world. About such creatures, I. J. Good remarked (in reference to an example in which H is the hypothesis that all crows are black, E is evidence reporting observation of a black crow, and $W(H:E)$ is a measure of the bearing of E on H):

Since the propositions H and E would be meaningless in the absence of empirical knowledge, it is difficult to decide whether $W(H:E)$ is necessarily positive. The closest I can get to giving $W(H:E)$ a practical significance is to imagine an infinitely intelligent newborn baby having built-in neural circuits enabling him to deal with formal logic, English syntax, and subjective probability. He might now argue, after defining a crow in detail, that it is initially extremely unlikely that there are any crows …

(Good 1968: 157)

David Lewis has referred to a creature of this sort as a *Superbaby*.[14] Fortunately, in this book we will not have much to do with such creatures. But it does seem clear that in a state of *complete* ignorance, the Superbaby would not have much basis—or rather, would not have any basis at all—for making those judgments of equipossibility "whose exact appreciation," as Laplace put it, "is one of the most delicate points of the theory of chance."

We should not expect the Superbaby to have precise credences in anything other than contradictions and logical truths. Precise, or even reasonably well-defined, credences are not bestowed upon us for free, but rather, are hard-won. The Superbaby's credal state would either be a reflection of innate prejudices with no rational basis, or else would be a tower made entirely of Good's wool, with indeterminate credences, second-order credences, etc.

2.3 The Principal Principle and its justification

So far, we have been talking about credence. The other notion of probability we are interested is the aleatory concept, which we will call *chance*. It is worth stressing that these are *not* rivals for the title of the One True Interpretation of Probability. They are distinct concepts, which serve distinct purposes.

The two concepts, chance and credence, are distinct, but, as Lewis (1980) has emphasized, they are connected. To make it clear that they are distinct, Poisson (in the quotation in the first section of this chapter) invites us to consider a situation in which we regard it as impossible that *Heads* and *Tails* have equal chances. This could happen if, for example the coin were known to have come from a source that produces coins that are either biased two-to-one in favour of Heads, or two-to-one in favour of Tails. If your credences are equally divided between these alternatives, then your credence that the coin will come up *Heads* is one-half, even though you are certain that the chance of *Heads* is *not* one-half.

Poisson is taking it for granted that, when you are uncertain about the chance of an event, your credence in that event is a weighted average of the

---

[14] Reported by Alan Hájek in "Staying Regular?" (Hájek, MS). As far as I know, Lewis never used the term in print.

chances you deem possible, weighted by your credences in the alternatives. If, for example, you are certain that the coin is either biassed two-to-one in favour of *Heads* (that is, the chance of *Heads* is 2/3), or two-to-one against *Heads* (chance of *Heads* 1/3), and your credences in these alternatives are $p_{2/3}$ and $p_{1/3}$, respectively, then your credence in heads should be the corresponding weighted average of these two alternatives for the chance of *Heads*.

$$Cr(H) = \frac{1}{3} \times p_{1/3} + \frac{2}{3} \times p_{2/3}. \tag{2.5}$$

If we vary $p_{1/3}$ and $p_{2/3}$, maintaining the condition that they add up to one, then $Cr(H)$ ranges between 1/3 (for $p_{1/3} = 1$) and 2/3 (for $p_{2/3} = 1$), and never ventures outside the interval $[1/3, 2/3]$.

We can rewrite (2.5) as,

$$Cr(H) = \frac{1}{3} \times Cr(ch(H) = 1/3) + \frac{2}{3} \times Cr(ch(H) = 2/3). \tag{2.6}$$

Even if you are not certain that the chance of *Heads* is one of these two alternatives, we can consider conditional credences, conditional on the supposition that it is. More generally, conditional on the supposition that the chance of heads lies within a certain subset Δ of the real line, your conditional credence, $Cr(H \mid ch(H) \in \Delta)$, should be a weighted average of the alternative chances within Δ. On this condition, $Cr(H \mid ch(H) \in \Delta)$ cannot be greater than the upper bound of elements of Δ, or lower than the lower bound.

One way to put this is that, for any interval Δ of the real line, and any proposition p, your conditional credence, conditional on the supposition that the chance of p is in Δ, must be in Δ.

$$Cr(p \mid ch(p) \in \Delta) \in \Delta. \tag{2.7}$$

This is one formulation of what Lewis (1980) dubbed the *Principal Principle*.[15] Typically, it is formulated in terms of propositions about exact, point-values of chances; in this formulation, we have

$$Cr(p \mid ch(p) = x) = x. \tag{2.8}$$

---

[15] It is also sometimes called *Miller's Principle*, after Miller (1966), who formulated it in order to argue that it led to a contradiction.

Formulation (2.8) can be thought of as a limit-case of (2.7). If $\{\Delta_n\}$ is a nested sequence of intervals whose sole common element is x, and if, for every n, you assign non-zero credence to the proposition $ch(p) \in \Delta_n$, then (2.7) entails

$$Cr(p \mid ch(p) \in \Delta_n) \to x \text{ as } n \to \infty. \tag{2.9}$$

The formulations (2.7) and (2.8) are not equivalent; (2.8) entails (2.7) but not *vice versa*. However, if (2.7) holds and the conditional probabilities invoked in (2.8) are defined, then (2.8) must hold for *almost all x*, that is, the set of all x, if any, for which it doesn't hold is a set to which you ascribe credence zero.

I prefer formulation (2.7), because any coherent probability function can assign positive probability to at most countably many point values of x in the unit interval, leaving uncountably many assigned zero probability, and I don't want to assume that conditional probabilities with zero-probability conditions are defined (for the reasons why, see Myrvold 2015). The chief import of the Principal Principle is that it turns frequency data into evidence about chances, as will be outlined in the next section, and, for that account of updating to go through, (2.7) suffices.

One can add to the conditioning propositions any proposition as long as it is, in Lewis' term, *admissible*, admissible propositions being those "whose impact on credence about outcomes comes entirely by way of credence about the chances of those outcomes" (Lewis 1980: 272). Adding this in, the principle is the condition that, for any interval Δ, and any admissible e,

$$Cr(p \mid e \,\&\, ch(p) \in \Delta) \in \Delta. \tag{2.10}$$

These versions of the Principal Principle (PP) involve only a single proposition whose chance is to be considered. In general, we will consider theories about chances that assign chance distributions to some algebra of propositions. To state the appropriate generalization of the PP, we need some terminology.

For any two probability functions P_1, P_2, for any w, $0 < w < 1$, the function

$$P_w = w P_1 + (1 - w) P_2 \tag{2.11}$$

is also a probability function, called a *mixture* of P_1 and P_2. A set of probability functions is *convex* if it contains all mixtures of all probability functions in the set; the *convex hull* of a set $\mathcal{P}$ of probability functions is the smallest convex set containing $\mathcal{P}$.

Let $\mathcal{A}$ be an algebra of propositions, and let $\mathcal{P}(\mathcal{A})$ be the set of all probability functions on $\mathcal{A}$. If $\mathcal{A}$ is finite, then any probability assignment on $\mathcal{A}$ can be represented as a finite list of real numbers. The set of all such assignments can, therefore, be represented as a subset of $\mathbb{R}^n$, and we can form a probability space by taking as our measurable sets the Borel subsets of this subset of $\mathbb{R}^n$. (See appendix for definitions.) For any measurable set Δ of probability assignments on $\mathcal{A}$, let $ch(\mathcal{A}) \in \Delta$ be the statement that the chance-function of $\mathcal{A}$ (that is, the probability function that represents the chances) is in Δ.

We generalize the idea that, conditional on the supposition that the chance-function is in a given set Δ, your credence in a proposition p must be a weighted average of the chances of p according to members of the set Δ. This can be expressed by the following Commandment, which we will take as our fullest version of the Principal Principle.

> Let $\mathcal{A}$ be an algebra of propositions, and let $\mathcal{P}(\mathcal{A})$ be the set of all probability functions on $\mathcal{A}$. For any Borel subset Δ of $\mathcal{P}(\mathcal{A})$, let $ch(\mathcal{A}) \in \Delta$ be the statement that the chance-function of $\mathcal{A}$ is in Δ. For any set Δ such that you have credences conditional on the supposition of $ch(\mathcal{A}) \in \Delta$, your conditional credence-function, conditional on $ch(\mathcal{A}) \in \Delta$ and any admissible evidence, should be in the convex hull of Δ.

Call this the Principal Commandment. This formulation does all the work that we need the Principal Principle to do, and is neutral with respect to the matter of whether credences conditional on propositions to which you ascribe zero credence are defined.

As Lewis (1994: 475–6) remarks, a special case of the Principal Principle is the case in which an agent's credences are entirely concentrated on a single point value for the chance of some proposition p, which somehow happens to be the actual chance of p. In such a case, if the agent's credences satisfy the PP, her credence in p will be equal to the chance of p. This is, in my estimation, an uninteresting case, because it involves an epistemic state that no reasonable agent is ever in, as complete certainty about a point value of a chance would be reasonable only if the agent had evidence that completely ruled out other values. Though (as we shall see in the next section) we can, given favourable circumstances, get *very good* evidence about the chance of some event, resulting in credences that are sharply peaked near the actual chance, complete certainty is not to be had.

The PP, by itself, dictates no relation between an agent's credences and the actual chances. In fact, satisfaction of the PP does not require there to *be* any actual chances; since it is a constraint on an agent's credences, all it requires is that the agent be able to formulate propositions about chances, to have credences in such propositions, and to have credences conditional on propositions about chances, whatever the actual truth values of those propositions may be. Even if there are in fact, no non-trivial chances in the world (that is, no chances other than one and zero), the Principal Principle still makes prescriptions about an agent's credence conditional on hypotheses about non-trivial chances.

This is worth mentioning, because one sometimes hears the special case that obtains when the agent knows for certain what the chances are as a gloss on the Principal Principle. That is, sometimes it is said that the PP requires one to set one's credences equal to the chances. If this were what it prescribed, it would seem somewhat mysterious—how could chances, which are features of the actual world, impose a normative constraint on an agent's credences?

Lewis said that the Principal Principle seemed to him to "capture all we know about chances" (1980: 266). This strikes me as a bit misleading, as the PP is a constraint on *credence about chances*. The PP doesn't really tell us anything about chances, not even whether there are any; as already mentioned, satisfaction of the PP by an agent's credences does not require that there be non-trivial chances, only that the agent can entertain hypotheses about them. It would be more apt to say that the PP captures all we know about credences about chances. Indeed, it seems to me that the PP is *constitutive* of the notion of *credence about chance*; we shouldn't regard an agent's credences to involve genuine credences about chances unless they satisfied the PP.

Consider the following dialogue. Alice and Bob are enjoying a day at the races. Equal odds are posted for Snowball and Excalibur.

ALICE: Which are you going to bet on?
BOB: I'm betting on Excalibur.
ALICE: Ah, so you think Excalibur has a better chance to win?
BOB: No, I think Snowball's got a better chance. That's why I'm betting on Excalibur.
ALICE: Don't you want to win?
BOB: Yes, of course I want to win.
ALICE: ???

Were such a dialogue to take place, it would be hard to escape the conclusion that Bob is confused, or not using words in the same sense that

Alice is, or misreporting his beliefs about chance, or that, despite what he says, there are other considerations than the prospect of winning his bet that are affecting his preferences. If you were to tell me that Bob could be thinking clearly, sincerely believe that Snowball has a better chance at winning than Excalibur, be motivated only by the prospect of winning his bet (and not, for example, out of some loyalty to Excalibur's owner, or the region from which Excalibur hails), and nonetheless prefer a bet on Excalibur to a bet at the same odds on Snowball, then, well, I honestly don't understand what you might mean by that.

From the facts that Bob wants to win his bet and that he prefers a bet on Excalibur to a bet on Snowball, we are licensed to infer that Bob's credence in the proposition that a bet on Excalibur will pay off is higher than his credence in the proposition that a bet on Snowball will pay off. We are thereby using his preferences and the utilities we ascribe to him as a guide to ascribing credences to him. Similarly, a condition on ascribing credences about chances to Bob is that his credences about the chance of an event p mesh with his credences about p in the way prescribed by the Principal Principle. If Bob were to be certain that Snowball has a better chance of winning—that is, if he ascribes non-zero credence only to chance-functions on which Snowball's chance of winning is higher than Excalibur's—then for Bob to place higher credence in a win by Excalibur than by Snowball would be a violation of the Principal Principle. If, therefore, he does place higher credence in Excalibur winning than Snowball, this places a constraint on the credences about chances that (barring confusion, etc.) we may attribute to him; it is not the case that he is certain that Snowball has a better chance of winning.

Construed in this way, as a constitutive principle for the notion of credence about chance, there isn't any mystery about why reasonable credences should be required to satisfy the Principal Principle.

2.4 Learning about chances

Suppose you have a coin whose bias you want to ascertain. The way to do it is to flip the coin a large number of times, and update your credences about the bias by conditionalizing on the evidence obtained.

Suppose that your prior credences about the chance of heads are represented by a density function f. That is, for any interval Δ,

$$Cr(ch(H) \in \Delta) = \int_{\Delta} f(x)\, dx. \qquad (2.12)$$

And let us suppose that you're not excessively dogmatic; for any interval Δ, you have non-zero credence that the chance of heads lies in that interval.

You flip the coin a large number of times, and tabulate the results. For example, consider a reduced experiment in which you stop at 10 flips, with the result

HTTHHHTHHT

Let E be the proposition that you obtained this result. Prior to the experiment, the chance of E, on the assumption that the chance of *Heads* on each individual toss is x, was[16]

$$ch(E \mid ch(H) = x) = x^6(1 - x)^4. \qquad (2.13)$$

For given E, this varies with x. Construed as a function of x, it is called the *likelihood function*.

$$l_E(x) = ch(E \mid ch(H) = x). \qquad (2.14)$$

To generalize: Suppose you toss the coin N times and obtain a result E containing m heads and $n = N - m$ tails, yielding an observed relative frequency of heads m/N. The chance of obtaining E, if the chance of heads on each trial is x, is

$$ch(E \mid ch(H) = x) = x^m(1 - x)^n. \qquad (2.15)$$

This is a function of x that takes its maximum value at the observed relative frequency m/N, and it is the more sharply peaked at that value, the larger N is. In Figure 2.1(a) this function is graphed for the example above, with $m = 6$ and $n = 4$; Figure 2.1(b) shows it for 1,000 tosses, with the same relative frequencies, that is, 600 Heads and 400 Tails.

---

[16] Note that, though this depends only on the number of *Heads* and the number of *Tails*, this is not the chance of obtaining 6 *Heads* and 4 *Tails*. To get the chance of obtaining 6 *Heads* and 4 *Tails*, you'd have to multiply by the number of ways to do that.

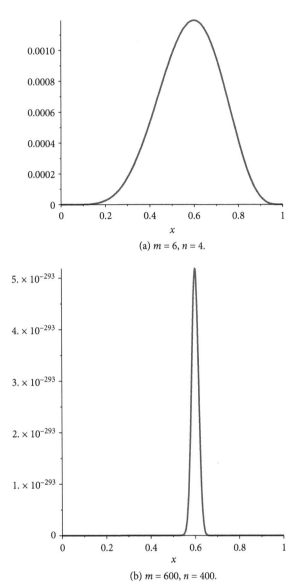

(a) $m = 6$, $n = 4$.

(b) $m = 600$, $n = 400$.

Figure 2.1. Likelihood functions

The Principal Principle requires that your credences satisfy the condition that, for any interval Δ,

$$Cr(E \ \& \ ch(H) \in \Delta) = \int_{\Delta} l_E(x)f(x) \, dx. \qquad (2.16)$$

(Recall, f is a density function representing your credences about the chance of *Heads*.) If, now, you update your credences by conditionalizing on the evidence E, this yields a new credence density function, f_E, which, because of (2.16), is proportional to the result of multiplying the old credence density function by the likelihood function.

$$f(x) \Rightarrow f_E(x) = l_E(x)f(x)/Cr(E). \qquad (2.17)$$

Because the likelihood function l_E is peaked at the observed relative frequency, conditionalizing on the evidence raises your credence function in regions near to the observed relative frequency and lowers it in regions far from it, and its effectiveness in doing so increases with the number of trials. Any initial credence represented by a density function that assigns a non-zero probability to every interval will, for sufficiently large N, be well approximated by one that has the shape of the likelihood function, normalized so that the total area under the curve is 1.

Because of the Weak Law of Large Numbers (see Appendix, §11.4), for large N the chance is high that the observed relative frequency will be close to the actual single-case chance of *Heads*. Thus, as long as your initial credences are not too dogmatic, the chance is high that your credence density function about the chance of *Heads* will end up sharply peaked near the actual chance. The Principal Principle plays an important role in this, by way of equation (2.16). The PP turns relative frequency data into evidence for updating one's credences about chances.

It bears emphasizing that conducting a test of this sort, and updating on the results, is not *guaranteed* to improve your credence about chance. For any possible sequence of outcomes of tosses, there is a non-zero chance of obtaining that sequence as a result of the experiment, and this includes sequences in which the observed relative frequency of heads is far from the actual chance. In such a case, the result will be misleading, and conditioning on the results will make your credences about the actual chance worse. Fortunately (and this is what makes the experiment worth doing), the chance of that happening is low.

2.5 Determinism, enemy of chance?

One way for there to be objective chances is for the fundamental laws of physics to have chances in them, that is, for the laws themselves to be stochastic laws. On a theory like that, the past history of the world up to a given time, together with the fundamental dynamical laws of nature, would not uniquely determine the course of future events, but instead bestow determinate probabilities upon future events.

But if one takes the laws of nature to be deterministic, perhaps taking this to be an *a priori* necessity, or if one takes all things to be predetermined by divine providence, there is a puzzle about what sense it might make to talk about objective chances. This is something that early authors on probability wrestled with. Some of them, on the basis of such considerations, declared probability to be wholly epistemic. Here's Jacob Bernoulli:

> In themselves and objectively, all things under the sun, which are, were, or will be, always have the highest certainty. This is evident concerning past and present things, since, by the very fact that they are or were, these things cannot not exist or not have existed. Nor should there be any doubt about future things, which in like manner, even if not by the necessity of some inevitable fate, nevertheless by divine foreknowledge and predetermination, cannot not be in the future. Unless, indeed, whatever will be will occur with certainty, it is not apparent how the praise of the highest Creator's omniscience and omnipotence can prevail.
>
> . . . . Seen in relation to us, the certainty of things is not the same for all things, but varies in many ways, increasing and decreasing. . . . *Probability*, indeed, is degree of certainty, and differs from the latter as part of the whole.
>
> (Bernoulli 2006: 315; translation of Bernoulli 1713: 210–11)

Though his official view is that probability is epistemic, the aleatory conception can be found in his work also, as Bernoulli speaks as if the equal probability of outcomes of throws of a fair die is a feature of the physical set-up.

> The originator of these games [that is, games of chance] took pains to make them equitable by arranging that the numbers of cases that result in profit and loss be definite and known and that all the cases happen equally easily [*ut casus hi omnes pari facilitate obtingere possent*]. . . . So, for example, the numbers of cases in dice are known; for a single die there are manifestly as

many cases as the die has faces. Moreover these all have equal tendencies to occur [*omnes aequè proclives*]; because of the similarity of the faces and the uniform weight of the die, there is no reason why one of the faces should be more prone to fall than another—as would be the case if the faces had dissimilar shapes or if a die were composed of heavier material in one part than another.

(Bernoulli 2006: 326–7; translation of Bernoulli 1713: 223–4)

Laplace (1814) echoes Bernoulli in all this. We find, in Laplace, a similar commitment to determinism, already quoted in Chapter 1. Probability, for Laplace, arises from our ignorance, not from the nature of things external to us. But there seems to be an objective aspect also to it; for Laplace, the probability of an event can be found by first identifying a partition consisting of mutually exclusive and jointly exhaustive events that are, in Laplace's phrase, *equally possible*, and, though "equally possible" cases are glossed as those about which we may be "equally undecided" (Laplace 1814: 7; 1902: 6), Laplace also speaks, in places, as if there is an objective matter of fact about which events are equally possible.

Writers on probability who followed in Laplace's footsteps were less ambiguous. One motivation (already seen in Bernoulli) for denying objective chance in the world is the charge of irreligion; if there is a deity who plans every detail of the unfolding of events, there can be no objective chance. We find Augustus de Morgan, in his *Essay on Probabilities*, defending the theory of probability against charges of this sort: "the word chance, in the acceptation of probability, refers to events of which the law or purpose is not *visible*." Furthermore, de Morgan insists,

The theory of probabilities absolutely requires, in its fundamental principles, the rejection of the notion that pure chance can produce any two events alike; that is, it presumes causation and order of some kind or other, that is, *providence* of some kind or other.   (de Morgan 1838: 24)

Following in his footsteps, Jevons writes,

Chance cannot be the subject of the theory, because there is really no such thing as chance, regarded as producing and governing events. This name signifies *falling* and the notion is continually used as a simile to express uncertainty, because we can seldom predict how a die, or a coin, or a leaf will fall, or when a bullet will hit the mark. But every one knows, on a little reflection, that it is in our knowledge the deficiency lies, not in the certainty of nature's laws. There is no doubt in lightning as to the point it shall strike; in the greatest storm there is nothing capricious; not a grain of sand lies upon the beach, but infinite knowledge would account for its lying there; and the course of every falling leaf is guided by the same principles of mechanics as rule the motions of the heavenly bodies.

Chance then exists not in nature, and cannot co-exist with knowledge; it is merely an expression for our ignorance of the causes in action, and our consequent inability to predict the result, or to bring it about infallibly. In nature the happening of a physical event has been pre-determined from the first fashioning of the universe. Probability belongs wholly to the mind . . . .

(Jevons 1874: 225)

2.6 The displacement of chance by frequency

There is a tension in these writers. Determinism seems to leave no room for objective chance. This suggests that probabilities are purely epistemic, having to do with our own degrees of belief, and not with the things themselves. On the other hand, there seem to be matters of fact about some chances, such as chance of *Heads* for a given coin, or perhaps even such things as the chance of succumbing to a given disease within a year, and these matters can be investigated empirically. There seems to be a need for a notion of objective chance that makes sense, even if the fundamental laws of physics are deterministic.

One strategy that has been tried is an appeal to frequencies. In his *Logic of Chance* (1866), Venn introduced a view that he proclaimed to be radically different from those of most other writers on the subject (p. xii). This is the view that has come to be known as *frequentism*.[17] Venn was strongly critical of the subjective conception of probability, represented, for him, by de Morgan. In addition, like so many others in the nineteenth century, Venn was struck by the phenomenon of aggregate order arising from individual disorder. The proper subject matter of probability, according to Venn, has to do with series of events that are individually unpredictable but which show tendencies towards stable frequencies.

---

[17] Ellis (1849) is often cited as a precursor of Venn's frequentism. Ellis suggests that "the true definition of probability" is "founded on a reference to the ratios developed in the long run" (§9), but the idea is not well developed in his brief essay.

We are now in a position to give a tolerably accurate definition of a phrase which we have frequently been obliged to employ, or incidentally to suggest, and of which the reader may have looked for a definition already, viz. the probability of an event, or what is equivalent to this, the chance of any given event happening. I consider that these terms presuppose a series; within the indefinitely numerous class which composes this series a smaller class is distinguished by the presence or absence of some attribute or attributes, as was fully illustrated and explained in a previous chapter. These larger and smaller classes respectively are commonly spoken of as instances of the 'event' and of 'its happening in a given particular way.' Adopting this phraseology, which with proper explanations is suitable enough, we may define the probability or chance (the terms are here regarded as synonymous) of the event happening in that particular way as the numerical fraction which represents the proportion between the two different classes in the long run.   (Venn 1866: §III.33: 106–7)

In considering the long run, it is essential that the events considered be taken to be of a fixed type. Though certain statistics, such as the annual number of accusations of witchcraft *per capita*, would exhibit fairly stable frequencies from year to year, these stable frequencies exhibit long-term changes (secular variation), as conditions change. To compensate for this, we substitute a hypothetical infinite series of events of the same type for the series that actually occurs in nature. The chance of an event is its limiting frequency in such a series (Venn 1866: 108).

Bernoulli, Laplace, de Morgan, Jevons, and others adopted a subjective conception of probability on the grounds that chance was inconsistent with determinism. Venn rejects subjective probability, and seeks a conception that does justice to the fact that chances are things that we learn about empirically, and that the sort of evidence we turn to, in estimating chances, is relative frequencies. This is a conception that is compatible with determinism, though compatibility with determinism was not Venn's primary motivation (see Venn 1888: §X.3). By the end of the nineteenth century, the aleatory conception of probability, that of single-case objective chance, found in Poisson and Cournot, had largely dropped out of discussions of probability. I conjecture that the primary reason for this was that objective chance was taken to be incompatible with determinism, which was taken for granted by most, and by many taken to be a truth known *a priori* or to follow from God's omniscience. Frequentism seemed to provide a viable substitute, providing objective probabilities without threat of conflict with determinism.

In the twentieth century, it became common to contrast the epistemic sense of "probability," not with the concept of chance, but with a frequency interpretation of probability (see e.g. Ramsey 1931: 157; Carnap 1945; 1950; Nagel 1955: 17–19). The concept of objective chance largely dropped out of the discussions. Absence of the notion of objective chance in so many discussions of the foundations of probability led Popper (1957; 1959) to conclude that he had an entirely new idea in single-case objective probabilities, which he called "propensities."[18]

2.7 Could probability be only credence?

There is a long history of avowals that probability simply *is* credence. We have already heard from Bernoulli, Laplace, de Morgan, and Jevons. In the twentieth century, prominent advocates of the identification of probability with credence have been Savage and de Finetti. Among physicists, it has vocal advocates in those who call themselves "QBists," for "Quantum Bayesians."

Arguments for a position of this sort fall into three classes. The first, which we have already seen, presumes determinism, and rejects chance as incompatible with determinism. A second sort involves a consideration, and rejection of, a limited array of candidate construals of objective probability. As already mentioned, by the twentieth century the concept of single-case chance had largely dropped out of discussion. Thus we find de Finetti rejecting objective notions of probability on the basis of a rejection of two conceptions, one that grounds probability in consideration of equally possible cases, and a frequency conception (see de Finetti 1989: §7, 10, from 1931, and de Finetti 1980: 71ff., from 1937: 16ff.). De Finetti is right to reject these conceptions, which will be the subject of our next chapter. Savage (1954: 4) identifies an objective conception of probability with a frequency interpretation, and takes it as a virtue of the subjectivist view that it is applicable to single cases. In a similar vein, QBists Fuchs and Stacey (2019) motivate their rejection of objective probability via an imaginary conversation with a "Pre-Bayesian" who advocates a frequentist conception of probability; no other conception of objective probability is considered.

---

[18] Hacking (1971: 343) rightly says, of Popper's propensity interpretation of probability, "Whatever name one likes to give it, this is the idea most current in the early days of probability…"

The weakest form of argument in favour of probability as exclusively epistemic presumes that the word "probability" is univocal and makes a case for accepting an epistemic notion, excluding the aleatory notion on the (usually tacit) premise that There Can Be Only One. This is what we find in, for example, de Finetti (1976).

These arguments in favour of an exclusively epistemic conception of probability are not very strong. Elimination of the aleatory conception would leave a gaping hole in the conceptual framework with which we operate. There are a host of situations in which we take the probability of some event as something about which there is a matter of fact, not dependent on any agent's state of mind. Roulette wheels for use in commercial casinos are carefully made to be free of bias that might lead to detectable patterns that customers might take advantage of. One can (and I presume that casinos do in fact do this) test a wheel for bias by subjecting data from a long series of trials to statistical analysis. Taken at face value, a procedure like that only makes sense if it is possible to entertain a hypothesis that there might be a bias in the wheel, unknown to anyone, and to subject that hypothesis to empirical test. One also tests hypotheses about chances outside of a gambling context, of course. Any analysis of experimental data that involves statistical analysis is, at its heart, aimed at estimating certain properties of a probability distribution from which the data is drawn.

Moreover, in statistical mechanics, probabilistic posits are routinely made, and predictions derived from them. From the Langevin equation one can derive conclusions about the probability distribution of the displacements of the particle over time, which, if confirmed, yield information about the probability distribution of the fluctuating force, which, in turn yields information about the processes that might be responsible for that force. Taken at face value, this procedure involves forming a hypothesis about a chance distribution in nature, and subjecting that hypothesis to experimental test.

Were we to be given a sufficiently strong argument that procedures such as these are not to be taken at face value, we might be motivated to try to find a way to make sense of them in other terms. Were we to be given an extremely powerful argument that no sense could be made of such procedures, it might become reasonable to reject them as ill-conceived. Fortunately, we are not even remotely close to being in such a position.

One way to make sense of such procedures would be to invoke chance in the fundamental laws of physics. But consider: the fact that one can detect a bias in a roulette wheel by statistical analysis of its behaviour is a fairly banal fact. The question of whether the fundamental laws of physics are deterministic or indeterministic is not banal in the same sense, and it may be one that is beyond our grasp. It had better not be the case that the banal observations we have made about the role of chance in our lives carries with it a commitment to indeterminism at the fundamental level.

So, there does seem to be an important role, in the way we deal with the world, for a notion of chance such that:

- The value of a chance does not depend on what anyone thinks of it, and, in fact, may be completely unknown to anyone.
- The value of a chance *does* depend on physical features of the chance set-up, and can be altered by a change in relevant physical features, whether or not anyone is aware of the change.
- We can formulate hypotheses about chance distributions, and subject such hypotheses to empirical test.
- Though frequency data is often useful in evaluating such chances empirically, the meaning of a claim about chance is not reducible to frequency claims, and can make sense in a single cases.
- We can explain why we should expect there to be statistical regularities of the sort that we do, in fact, observe.
- Employment of the notion of chance in these ways does not require a commitment to indeterminism in the fundamental laws of physics, and can make sense in the context of deterministic physics.

The reader may expect that what I will offer is a notion of objective chance that fulfills these *desiderata*. That is *not* what I am going to do! I think that, in order to formulate a concept of probability that meets these demands, we need to go beyond the familiar dichotomy of *objective chance* and *epistemic credence*. What I will offer is a hybrid concept that combines both epistemic and physical considerations. We will meet this concept, which I call *epistemic chance*, in Chapter 5.

3

Two Non-senses of "Probability"

3.1 The insufficiency of indifference

There is a temptation to think that probability can be defined in terms of mere counting, that is, to take the probability of an event to be just the ratio between the number of ways the event can occur and the number of ways the world could be. Associated with this is a temptation to believe that there is a principle, widely known as the "principle of want of sufficient reason," or "principle of insufficient reason," until Keynes (1921) renamed it the "Principle of Indifference," which can generate probabilities out of ignorance. It has been called "The Basic Assumption of Statistical Mechanics" (Jackson 1968: 83) and one well-known author has said that, in statistical mechanics, "the Principle of Indifference is basically the best we can do" (Carroll 2010: 168). On the other hand, it has long been roundly criticized in the literature on probability, and one philosopher has called it "that notorious dead horse of the philosophy of probability" (Butterfield 1996: 212). At risk of incurring the wrath of the Society for the Prevention of Cruelty to Deceased Equines, we need to discuss the principle, because understanding why it's insufficient will help us understand what is needed from our concept of probability, and why probability cannot be a simple matter of counting possibilities.

3.1.1 D'Alembert's folly

Much of what goes on in an elementary course of probability, and much of the early development of what we now call probability theory, consists of counting possibilities. Consider, for example a problem posed to Pascal by the Chevalier de Méré:[1] Suppose a pair of fair dice are to be tossed 24 times. Is it better to bet in favour of the proposition that at least one of the 24 tosses will be a pair of sixes, or against it?

---

[1] Pascal mentions this problem in a letter to Fermat of July 29, 1654. See Ore (1960: 411–13) for discussion.

Beyond Chance and Credence. Wayne C. Myrvold, Oxford University Press (2021).
© Wayne C. Myrvold.
DOI: 10.1093/oso/9780198865094.003.0003

The answer consists, in effect, of counting the number of ways that the 24 tosses can turn out, then counting the number of outcomes that include a pair of sixes on some toss, and the number that don't. We don't actually list the outcomes, of course; the number is astronomically large. But, if you work it out, there are

$$22,452,257,707,354,557,240,087,211,123,792,674,816$$

ways that a sequence of 24 tosses can turn out, of which there are

$$11,033,126,465,283,976,852,912,127,963,392,284,191$$

sequences in which a pair of sixes at least once, and

$$11,419,131,242,070,580,387,175,083,160,400,390,625$$

sequences in which a pair of sixes never occurs. So, if each of these possible outcomes is equiprobable, it is marginally better to bet that a pair of sixes won't occur. Much of the mathematical ingenuity of early work in probability theory consists of development of methods for handling this sort of combinatorics.

This procedure suggests that we can simply define the probability of winning a bet as the proportion, out of all the possible cases, of cases in which you win. Indeed, this is how Laplace defined probability, in his seminal *Philosophical Essay on Probabilities*, as his First Principle of the calculus of probabilities.

> *First Principle.*—The first of these principles is the definition itself of probability, which, as has been seen, is the ratio of the number of favorable cases to that of all the cases possible.
>
>   (Laplace 1902: 11, from Laplace 1814: 12).

Things are not so simple, however, and this does not suffice as a definition of probability. The reason for its inadequacy can be illustrated by a game considered by d'Alembert (1754). Your goal is to get *Heads* in a coin toss, in at most two tosses. A coin is tossed. If it comes up *Heads*, you win. If it comes up *Tails*, you get to try again. If it's *Heads* on the second toss, you win; if *Tails*, you lose. What are the odds that you will win?

There are three ways the game could go.

1. *Heads*, first toss.
2. *Tails*, first toss; *Heads*, second toss.
3. *Tails*, first toss; *Tails*, second toss.

In two of these cases, you win; in one, you lose. Now, if you *seriously* thought that Laplace's First Principle could serve as a definition of probability, then you would conclude (as did d'Alembert) that the probability of winning is 2/3, instead of 3/4, the correct answer that d'Alembert derides.

This has, rightly, been regarded as serious stumble on d'Alembert's part; a misstep taken by an otherwise brilliant mind. But we should be clear about what his error was. His mistake was to take seriously what many at the time were saying, and, regrettably, what even today many continue to say—namely, that we can define the probability of *A* as the ratio of the number of possible cases in which *A* occurs to the total number of possible cases.

Not only does application of Laplace's First Principle give the wrong answer in cases like these; indiscriminate application of the Principle could readily lead to incoherence. Suppose someone else, watching the game, bets on whether the first toss is *Heads*. That person's game is over on the first toss, and there are only two cases: *Heads* or *Tails*. Are there even odds of winning that game?

Laplace knew, of course, of cases like this, and that is why he *didn't* regard his First Principle as an adequate definition of probability. Having advanced an inadequate definition of probability in his First Principle, he proceeds to supply the needed *caveat* in his Second.

> *Second Principle.*—But that supposes the various cases equally possible. If they are not so, we will determine first their respective possibilities, whose exact appreciation is one of the most delicate points of the theory of chance.

There is a difference between d'Alembert's three cases. The first consists of only one toss; the second two agree on the outcome of the first toss and are distinguished by the outcome of the second. They are not, in Laplace's phrase, "equally possible."[2]

---

[2] In his earlier exposition of 1773, there is no separation of the two principles. What he says there is,

But what does "equally possible" mean? Can some things be more possible than others? If it means the same as equally probable, then Laplace's supposed definition is flatly circular.

Laplace seems to have meant something different (see Hacking 1971), but precisely what one *could* mean remains a delicate point. We are familiar with the fact that some things are harder to achieve than others. It is harder to flip a coin so that it lands and stays balanced on its edge (on a hard, smooth surface) than it is to flip it so that it lands on a face, so much so that we usually disregard the possibility of landing on an edge when calculating the odds of a coin toss. This isn't merely a matter of idiosyncratic limitation; it would not be easy to design a machine that would reliably do this, and we would not expect there to be a natural process that could reliably land a coin on its edge. That is, if the coin is subject to the usual influences, one could construct setups involving magnetized coins and a magnetic field that tended to align the coin vertically. The difficulty of landing a coin on its edge has to do, not merely with *number of cases*, but with the physics of the coin, the dynamics governing how it behaves.

This suggests a refinement of the evaluation of probabilities by counting cases. Instead of counting cases individuated at the level of macroscopic events, such as the outcome of the coin toss, one looks at the possibilities of microscopic initial conditions. We want to say that, in an ordinary coin toss, there are "more" initial conditions of the coin that lead to it landing on a face than to its landing on its edge. In the presence of a magnetic field tending to align the coin vertically, more of the initial conditions will lead to its landing on its edge. There's something right about this, and what is right (and what isn't) will be the subject of much of this book. But we shouldn't fool ourselves into thinking that things are going to be simple. Crucially, we're still not going to get a *definition* of probability from it, as the move from macroscopic to microscopic does nothing to alleviate the problem with taking Laplace's First Principle as a definition of probability. A judgment of which cases are (at least approximately) equiprobable is required, and the

The probability of the occurrence of an event is thus just the ratio of the number of favourable cases to that of all possible cases, when we see no reason why one of these cases would occur rather than another.

La probabilité de l'existence d'un événement n'est ainsi que le rapport du nombre de cas favorables à celui de tous les cas possibles, lorsque nous ne voyons d'ailleurs aucune raison pour laquelle l'un de ces cas arriverait de plutôt que l'autre.

(Laplace 1776, §XXV, in *Oeuvres Complètes*, Vol. 8, p. 145)

relevant sense of equiprobability won't have to do with merely counting cases or a geometrical evaluation of a volume of state space; the dynamics of the system will be have to come into play. More on this in Chapter 4.

3.1.2 The Principle of Indifference and its discontents

In the previous chapter, in §2.2.4, we mentioned a temptation to conflate two very different things: an absence of any known reason to judge one alternative more probable than another (which would warrant at best a suspension of judgment), and a positive judgment of equiprobability. Conflation of this sort is frequently found in discussions of what has been called the Principle of Want of Sufficient Reason.

In this vein, we find de Morgan writing, in connection with an example involving a lottery,

> Why does the lottery ticket of the preceding instance bear the character of being exactly worth 1*l*? Not as any consequence of the accuracy of the preceding process, supposing it accurate, but because we do not know why we should exceed rather than fall short of it. It appears to me that many of our conclusions are derived from this principle, which is called in mathematics *the want of sufficient reason*. A ball is equally struck in two different directions, the table being uniform throughout. In what direction will it move? In the direction which is exactly between those of the blows. Why? No positive reason is assignable (experiment being excluded); but from the complete similarity of all circumstances on one side and the other of the bisecting direction, it is impossible to frame an argument for the ball going more towards the direction of one blow, which cannot immediately be made equally forcible in favour of the other. The conclusion remains, then, balanced between an infinity of possible arguments, of which we can only see that each has its counterpoise. Now whether we adopt the above conclusion as to probability for its exactness or for its want of demonstrable inexactness one way more than the other, it is still a principle of human action, and as such is adopted. Many writers on probability speak of it as being a maxim which, if it were not adopted, ought to be.
>
> (de Morgan 1838: 10)

Though de Morgan distinguishes two ways in which lottery tickets might be deemed as having the same value—one, a known accuracy of the process by which the tickets are drawn, and the other, ignorance of any reason to prefer one to the other—he slides between the two without sufficient notice of their difference. He invokes an analogy with symmetry arguments in physics. If it is known that a physical set-up has a certain symmetry, we can draw conclusions from this symmetry about its behaviour. But this is different from simply being unaware of any relevant asymmetries!

We find another statement in Jevons (1874: 228).

> The calculation of probabilities is really founded, as I conceive, upon the principle of reasoning set forth in preceding chapters. We must treat equals equally, and what we know of one case may be affirmed of every case resembling it in the necessary circumstances. The theory consists in putting similar cases on a par, and distributing equally among them whatever knowledge we possess. Throw a penny into the air, and consider what we know with regard to its way of falling. We know that it will certainly fall upon a side, so that either head or tail will be uppermost; but as to whether it will be head or tail, our knowledge is equally divided. Whatever we know concerning head, we know also concerning tail, so that we have no reason for expecting one more than the other. The least predominance of belief to either side would be irrational; it would consist in treating unequally things of which our knowledge is equal.

Jevons' gloss on the Principle, as the injunction to "treat equals equally," highlights the reason, already seen clearly by Laplace, why the Principle by itself does not yield any probabilities: without a specification of *which* alternatives are to be treated as equals, it is utterly empty. And, if we are already in possession of a partition consisting of propositions that we judge to be equiprobable, no Principle is needed to tell us to treat them as equiprobable.

The Principle drew sharp criticism from a number of writers on probability, notably Venn (1866: §IX.4), von Kries (1886), and, influenced by von Kries, Keynes (1921). Keynes renamed the Principle the "Principle of Indifference," and formulated it thusly.

> The Principle of Indifference asserts that if there is no *known* reason for predicating of our subject one rather than another of several alternatives, then relatively to such knowledge the assertions of each of these alternatives have an *equal* probability. (Keynes 1921: 42)

Keynes points out that the Principle doesn't get us off the ground until we form a judgment of which set of alternatives it is to be applied to. Of a book whose colour is unknown, should we assign equal probability to its being red and not red? Or should we assign equal probability to each of red, black, and blue? It doesn't help to tell us to treat equals equally; which alternatives are to be regarded as equal?

The principle gets into deeper trouble when applied to infinite sets of exclusive alternatives. Suppose you are told that, though the universe is infinite in extent, there are only finitely many stars, and you know nothing else. What should your credences be about *how many* stars? Since probabilities of exclusive alternatives are additive, and they must be less than one, there is no way to assign an equal non-zero probability to every statement of the form "There are n stars."

It is sometimes said that, for probabilities involving quantities that can take on any value from a continuum (say, an interval of the real line), the problem becomes worse. But really, it's just the same problem over again. For such quantities, the Indifference Principle presumably enjoins one to adopt a uniform probability distribution, represented by a density function that is constant. But this is just the same as saying: For any division of the range of values that the quantity can take on into n equal segments, assign equal probability to each. But then this requires an answer to the question: what is to be counted as a division into n *equal* segments?

To take a simple example: you know that number x is greater than zero and less than one. What should your degree of belief be that it is less than $1/2$? It might seem obvious that the Principle of Indifference dictates that the answer be $1/2$, but consider: If x is greater than zero and less than one, so is x^2. Shouldn't the Principle of Indifference enjoin us to assign equal credence to x^2 being above and below $1/2$?

But we can't have both. On the assumption that x is greater than zero and less than one, x is less than $1/2$ if and only if x^2 is less than $1/4$. A credence-assignment that assigns equal credence to x^2 being in equal intervals assigns credence $1/4$ to x^2 being less than or equal to $1/4$, not $1/2$.

The fact that an injunction to adopt a uniform probability distribution requires supplementation by a specification of which variables the distribution is to be uniform in will come up in our discussion of statistical mechanics. Invocations of a Principle of Indifference in statistical mechanics assume, either explicitly or implicitly, a particular choice of parameterization, namely, that we parameterize the state space of the system by what are known as canonical variables. For a system of n particles, this would consist of specifying the position and momentum of each particle. This is not the only possible choice of parameterization, and imposing uniformity with respect to different parameterizations yields different probability distributions over subsets of the state space.

It turns out that, in the case of statistical mechanics, the dynamics of the system will help us out. If we are trying to model a system that has relaxed to equilibrium, and if we think that this should be represented by a probability distribution that is stable in time, then this restricts the possible candidates. In some non-equilibrium situations, dynamics will provide a clue as to an appropriate probability distribution. But this leads us away from the Principle of Indifference, which is essentially static, to a conception that bases probabilities on considerations about how the system behaves.

3.1.3 Notes on random chords: some prehistory of the so-called Bertrand "Paradox"

A recommendation to adopt a uniform probability distribution over a continuous space of possibilities fails to specify any distribution whatsoever, until it is supplemented by a choice of variable that is to be taken to be uniformly distributed. Examples illustrating this point are often associated with the name of Joseph Bertrand, whose *Calcul des Probabilités* (1889) opens with a series of examples intended to illustrate the by-then familiar point that a purported definition of probability in terms of ratio of number of favourable cases to all cases possible must be supplemented by the condition that the cases are "equally possible."[3]

The most famous of these involves a chord of a circle chosen at random, and asks for the probability that the chord is smaller than the side of an equilateral triangle inscribed in the circle. Bertrand suggests three ways of construing the question. The first involves choosing one endpoint at random,

---

[3] Bertrand, in fact, echoes Laplace on this.

> *The probability of an event is the ratio of the number of favorable cases to the total number of possible cases.* A condition is implicit: all the cases have to be equally possible. The definition makes no sense without this restriction.

> "*La probabilité d'un événement est le rapport du nombre des cas favorables au nombre total des cas possibles. Une condition est sous-entendue: tous les cas doivent être également possibles. La définition, sans cette restriction, n'autait aucun sens.*"

(Bertrand 1889: 2)

and then choosing the direction of the chord, with uniform distribution over all possible angles between the chord and the tangent to the circle. This yields a probability of 1/3. The second involves fixing the direction, and then imposing a uniform distribution over distances from the diameter. This yields an answer of 1/2. The third way of disambiguating the question involves choosing the midpoint of the chord at random within the interior of the circle, with the probability that it falls within a given region proportional to the area of the region. This yields an answer of 1/4. What Bertrand says about this is:[4]

> Among these three responses, which is the true one? None of the three is wrong, none is right, the question is ill-posed.   (Bertrand 1889: 5)

Bertrand is, of course, right about this. This is true even if, given a physical realization of a chance set-up that can be thought of as choosing a chord at random in a circle, there are physical considerations that can be employed to pick out one probability distribution as the one appropriate to that set-up.[5]

If one thought that there was a Principle of Indifference that must specify a unique answer to this question, one would find the need for disambiguation paradoxical (though it is clear that Bertrand himself does not), and, indeed, the fact that the bare statements of problems such as those considered by Bertrand do not suffice to uniquely determine their answers has (regrettably) come to be referred to as *Bertrand's paradox.*[6]

---

[4] "Entre ces trois réponses, quelle est la véritable? Aucune des trois n'est fausse, aucune n'est exacte, la question est mal posée."

[5] See Jaynes (1973) for one disambiguation of the problem, with an analysis that invokes dynamical symmetries to pick out a unique solution.

[6] It is not clear to me who was the first to treat Bertrand's example as a paradox, rather than (as Bertrand saw it) an illustration of a familiar point. The widespread currency of the phrase probably stems from Poincaré's *Calcul des Probabilités* (1896: 94; 1912: 118), first published in 1896, but based on lectures delivered in 1893–4. Though Poincaré speaks of a paradox, he is clear that there really isn't one. After outlining two different calculations of the asked-for probability, he writes, "Why this contradiction? We have made different hypotheses in the two cases; we have defined the probability in two different ways" (Poincaré 1896: 96; 1912: 120). ("Pourquoi cette contradiction? Nous avons fait des hypothèses différentes dans les deux cas, nous avons défini la probabilité de deux manières différentès.")

The first published reference to a Bertrand paradox that I have been able to find occurs in a report on the first summer meeting of the American Mathematical Society, held in Aug. 1894, which speaks of "Bertrand's renowned paradox," in connection with a talk by George Bruce Halsted entitled "Bertrand's paradox and the non-Euclidean geometry" (Fiske 1894: 5). Halsted's name is best known these days as translator of Poincaré's popular books into English, and, as this talk occurred shortly after the lectures on which Poincaré's book is based, one wonders whether he was in communication with Poincaré at the time.

Bertrand was not the first to make the point, nor was he the first to illustrate it by way of a choice of chords of a circle. Problems involving random chords had, in fact, been the subject of a lively and spirited debate in the pages of *The Educational Times*, two decades earlier.[7]

The discussion began with a dispute over the solution of what was called the "four-point problem": Given four points chosen at random on an infinite plane, what is the chance of one of them falling within the triangle formed by the other three (Wilson 1866a)? Discordant solutions were given, and there was a dispute over which was the correct answer (Ingleby 1866a; 1866b; Whitworth 1866; de Morgan 1866; Woolhouse 1866a). One Hugh Godfray weighed in on the dispute, offering what to our ears sound like words of wisdom:

> I believe it will be found that the discordance arises from the fact, that the word *random* is not sufficiently defined in the question; and the possibility of considering it in different ways, makes so many different problems, of which the various results are solutions.   (Godfray 1866: 73)

He proceeds to illustrate the point in connection with a question that had been posed a decade earlier by W. S. B. Woolhouse:[8] "Two chords are drawn at random in a circle, what is the chance that they will intersect?" There are, Godfray says, at least three ways in which the question may be interpreted. We could choose two points, independently, with uniform probability on the circle, and form a chord by joining them. Or else one could envisage a process in which the distance of the midpoint of the chord from the center of the circles is chosen from a uniform distribution between 0 and the radius, or one in which the length of the chord is uniformly distributed, between 0 and the diameter. Once the problem is duly specified, the question becomes a

---

[7] This was a periodical in which mathematical problems were posed and solved; there was also a bi-annual reprint (with supplements) issued under the title *Mathematical Questions, with their Solutions, from the "Educational Times," with many Papers and Solutions not published in the "Educational Times"*. See Grattan-Guinness (1992) for more on these periodicals.

[8] Woolhouse was the editor of *The Lady's and Gentleman's Diary*, a magazine of poetry, puzzles, and recreational mathematics. Question 1904 in no. 153 of the *Diary* (1856, p. 71), reads,

> In a dark room, two persons each of them draws a chord at random across a circular slate: what is the chance that they will intersect?

Two solutions, one by a Dr. Rutherford and one by W. J. Miller, editor of *Mathematical Questions*, were published in the following number (Rutherford 1857; Miller 1857). The generalization to an arbitrary number of chords was posed and solved by Woolhouse as Question 1894 of *Mathematical Questions*, vol. V (pp. 110–20), 1866.

well-posed question with a unique solution. Godfray's diagnosis of the dispute over the four-point problem is along the same lines: those offering different solutions are, in effect, solving different problems, as each has disambiguated the original, ambiguous question differently.

From the modern perspective, this is obviously right. However, this insight has been hard-won, and a symptom of this is the push-back that Godfray received. Woolhouse insisted that the interpretation of the question adopted in his own solution of question 1894 is "the only true meaning" (Woolhouse 1866b: 81). J. M. Wilson objected to Godfray's talk of various laws according to which something may be chosen at random; to draw a line at random is to choose "according to no law"; "to speak of random lines drawn according to a special law, with Mr. Godfray, is to me unintelligible" (Wilson 1866b: 82).

In his reply to Wilson, Godfray (1867) shrewdly pointed out that, though Wilson professed to find no sense in talking about random lines chosen according to this or that law, in the immediately preceding paragraph Wilson himself had offered a precise definition of what is to be understood by a random point, namely, "that it is equally likely to fall into any of the small equal areas into which we suppose space to be divided." Godfray urged the same precision in specifying what one means by a random line.

Not all were convinced; this note of Godfray's garnered replies from Woolhouse (1867; 1868) and from M. W. Crofton (1867). Crofton joined the others in denying that there is any ambiguity in the use of the phrase "at random," either in common usage or mathematical discourse.

> The expression "*at random*" has in common language a very clear and definite meaning; one which cannot be better conveyed than by Mr. Wilson's definition,—"*according to no law*." It is thus of very wide application, being often used in cases altogether beyond the province of mathematical measurement or calculation.   (Crofton 1867: 84; cf. Crofton 1868: 182)

Crofton's note in the *Educational Times* on this topic reappeared as expository paragraphs in an article published the following year in the *Philosophical Transactions of the Royal Society*, giving some results in geometrical probability, or, as Crofton called it, *local probability* (Crofton 1868).

What seems to be going on, in Wilson's and Crofton's initial replies to Godfray, is a conflation similar to the one we have already drawn attention to, in connection with the Principle of Indifference. There, an absence of judgment was being conflated with a judgment of equiprobability. Here, it seems that saying nothing about the manner by which a point or line is chosen is being conflated with a specification of a particular probability distribution, namely, a uniform one. Confusion then arises from the fact that differing probability distributions are obtained, depending on which parameter or parameters are to be taken to be uniformly distributed, and different choices seem to different people the natural, or obvious choice.[9]

Interestingly, though Bertrand's take on the problem is essentially the same of that of Godfray—namely, that questions of this sort require disambiguation as to which probability distribution is being invoked—it is possible that Crofton served as the conduit of the chord example from the pages of the *Educational Times* to Bertrand. Crofton's paper on local probability, which includes examples involving lines drawn across circles, was known to Bertrand, who in 1870 made reference to a theorem contained therein (Bertrand 1870: §506).

The dispute re-erupted in the pages of the *Educational Times* a decade later, in connection with a solution, by Elizabeth Blackwood, of a related question (vol. 28, question 5461) about random chords of a circle (Blackwood 1877; Woolhouse 1877). This time the disputants were moved to take the discussion beyond the limits of dry academic prose; Blackwood's solution prompted a criticism in comic verse from Helen Thomson (Thomson, 1878), parodying Walter Scott's "The Lady of the Lake." This prompted a longer verse reply from Blackwood (1878), which was supplemented by a more sober prose exposition of Blackwood's point of view, followed by a plea from the editor,

> We shall be glad if our correspondents will henceforth stick to prose, as, for subjects such as these, verse is wholly unsuitable.

The dispute in the *Educational Times* was summarized in the reprint volume in a note by the editor, W. J. C. Miller, appropriately entitled "Notes on Random Chords" (Miller 1878). In this note, Miller quotes from correspondence with Crofton, who, in spite of his insistence a decade earlier on the unequivocal sense of "at random," now acknowledges that problems of this sort can be disambiguated in various ways, and the answer will depend on how the problem is construed.

---

[9] It is perhaps worth emphasizing, once again, that, even in cases in which there is one choice that seems natural to everyone, this does not obviate the fact that a choice must be made, in order to specify a probability distribution!

Of course the results will depend on the manner in which we suppose "a random chord" of the area to be drawn. The results obtained by Miss BLACKWOOD and others are quite correct on the supposition that two points are taken on the perimeter of the circle (or any other convex figure) at random, and joined. I would observe, however, that it would be just as natural to take any two points at random within the circle and join them; but the result would be quite different; the problem indeed would, in this form, be a very difficult one, and would be an interesting exercise for our contributors.   (Crofton, quoted by Miller 1878: 18)

A chapter in which such matters are discussed was added by Venn to the third edition of *The Logic of Chance*. Venn endorses the now-standard view,

in these, as in every similar case, we always encounter, under this conception of 'randomness', at some stage or other, this postulate of ultimate uniformity of distribution over some assigned magnitude: either time; or space, linear, superficial, or solid. But the selection of the stage at which this is to be applied may give rise to considerable difficulty, and even arbitrariness of choice.   (Venn 1888: 100)

3.1.4 The accuracy-based argument for indifference

Richard Pettigrew (2016a; 2016b) has recently presented an argument, based on considerations of epistemic accuracy, for the Principle of Indifference. The argument is intended to show that a Superbaby[10] who considers only the algebra of propositions generated by a finite partition is constrained by rationality considerations to attach equal credence to all members of the partition.

Pettigrew's justification does not evade the criticisms of the Principle of Indifference we've already discussed. In order to be applicable, it requires as input a partition that has antecedently been judged to consist of propositions that are to be treated on a par. As it might not be obvious that Pettigrew's argument does require this, it's worth taking a look at how the argument works, to see where the requirement comes in.

The argument is based on the idea that the epistemic value of a credence function is to be judged by way of measuring its closeness to the truth,

---

[10] Refer back to §2.2.4 if you've forgotten what a Superbaby is.

or *accuracy*, a concept that we have already met in §2.2.1. The version of the Principle of Indifference that the argument from accuracy is meant to underwrite is the following:

> A Superbaby who, in a state of complete ignorance, considers a finite partition $\mathcal{F}$, ought to ascribe equal credence to each member of the partition.

Note that this formulation of the Principle embraces the partition-relativity that Keynes complained about. Suppose one Superbaby contemplates a partition $\{P_1, P_2\}$, and another subdivides P_2 into alternatives Q_1 and Q_2, and considers the partition $\{P_1, Q_1, Q_2\}$. Then Pettigrew's Principle of Indifference obliges the first Superbaby to place credence $1/2$ in the proposition P_1, and the second Superbaby to accord the same proposition credence $1/3$. This, as Pettigrew notes, does not lead to outright contradiction. But it places the Superbaby under an odd combination of freedom and constraint. Given a partition, the Principle is meant to uniquely constrain the Superbaby's credences. But the Superbaby, presumably, is free to choose which finite set of propositions it considers. This means that, for any proposition P, and any rational number p between 0 and 1 that the Baby might desire to take as its credence in P, the Baby may constrain itself to have credence p in the proposition P, by embedding it into an appropriate partition. I find it odd that a supposed constraint of rationality should leave the Baby open to ascribe to a given proposition any credence it wants (or, at least, a credence arbitrarily close to the credence it wants), and it leaves me wondering what the force of the constraint is. This oddity does not amount to outright inconsistency, and a proponent of the Principle might not regard it as objectionable. Indeed, Pettigrew declares this context-dependence to be "entirely appropriate" (2016a: 57) and "exactly right" (2016b: 167).

Pettigrew's argument for the Principle of Indifference proceeds from the premise that a Superbaby establishing its initial credences should adopt a Maximin principle. Given a candidate initial credence function, the Superbaby should consider the worst case, that is, the eventuality in which that credence function has the worst accuracy, and seek to maximize this worst-case accuracy. It is easy to see that, on the Brier score, there is a trade-off. I can increase accuracy in one eventuality only at the expense of decreasing it in another. The credence function that maximizes the worst-case accuracy is the uniform one that assigns the same credence to all elements of the partition under consideration.

Though the Maximin condition has a certain ring of plausibility, the argument falls afoul of the traditional sources of discontent with the Principle of Indifference. For one thing, the argument presupposes that the Superbaby has precise credences, rather than a credal state that is made entirely of Good's wool. Even if we grant the Superbaby precise credences, reasonableness of the Maximin principle requires a prior judgment that the alternatives considered are to be regarded as being on a par epistemically. To see this, consider the following partition. Let H_1 be a proposition that details the entire macroscopic history of our universe, up to some time-slice containing here and now. Let H_2 be the proposition that, ten minutes from now, Jupiter will spontaneously divide into exactly 1,000 blobs which will then arrange themselves into a pattern that spells out, as seen from Earth, the words "Don't Panic." Consider the four-element partition (using overbar for negation), $\{H_1 \& H_2, H_1 \& \bar{H}_2, \bar{H}_1 \& H_2, \bar{H}_1 \& \bar{H}_2\}$.[11]

Consider, now, two Superbabies, Arthur and Trillian. Arthur is very cautious, and wishes to maximize worst-case accuracy. He, therefore, accords each element of the partition equal credence. Trillian, on the other hand, regards the first member of the partition as very implausible, and accords it very small credence, a minuscule fraction of the credence she accords to the second element.

These prior credences yield very different conditional credences. Consider, for example, the conditional credence $Cr(H_2 \mid H_1)$. Arthur's priors assign the value of 1/2 to this. Trillian's priors assign it a very low value.

They are both taking an epistemic risk. If $H_1 \& H_2$ turns out to be the true element of the partition, then Arthur's accuracy is higher than Trillian's. If, however, $H_1 \& \bar{H}_2$ turns out to be true, Trillian's accuracy is higher than Arthur's. Arthur's worst-case accuracy is better than Trillian's, but Trillian's best-case accuracy is better than Arthur's.

Can Arthur convict Trillian of irrationality? He may point out that she's taking a risk of doing worse than him, if $H_1 \& H_2$ turns out to be true. But she might regard the possibility of this occurring as so remote that the prospect of improved accuracy if it *doesn't* occur is worth the gamble. That is, because she regards $H_1 \& H_2$ as less plausible than some of the alternatives, she regards the trade-off as worthwhile: she gets improved accuracy if it's not true. Arthur will be able to persuade her that the risk isn't worth taking only if he can

---

[11] As far as I can tell, Pettigrew's argument is meant to apply to any partition whatsoever, whatever the set of propositions entertained by the Superbaby happens to be, and so such a partition is meant to fall within the scope of the argument.

convince her that she ought to treat each element of the partition as an equally plausible alternative. That is, she should regard arguments based on the Maximin principle to have force only if she has antecedently judged the alternatives to be epistemically on a par.

If Trillian has credences different from Arthur's, or even a qualitative judgment that some elements of the partition are less plausible than others, there is nothing that Arthur can say that will persuade her to drop that judgment and adopt his Maximin-based credences. This is because Trillian will judge such a move to involve a drop in expected accuracy. Thus, the Maximin rule, unlike, say, an injunction to avoid a dutch book, cannot be utilized as a regulative principle to correct credences that it judges to be faulty.

Pettigrew acknowledges that there are no arguments that Arthur could use to persuade another agent who does not share his extreme epistemic conservatism (2016a: 46; 2016b: 167).

> At this point, it seems to me, we have reached normative bedrock: one cannot argue for cognitive conservatism from more basic principles. Thus, to those who reject cognitive conservatism... I can only recommend the argument of this paper as an argument for the following (subjunctive) conditional: Cognitive Conservatism ⇒ Principle of Indifference.
>
> > (2016a: 46; see also 2016b: 166–7)

The conditional *Cognitive Conservatism ⇒ Principle of Indifference* is, indeed, a theorem. Whether the conditional is to be used in an application of *modus ponens* or *modus tollens* is left to the judgment of the reader (but see also Myrvold 2019a).

3.1.5 The Principle of Maximum Entropy

E. T. Jaynes (1957a; 1957b) has proposed a generalization of the Principle of Indifference, which he calls the "Principle of Maximum Entropy" and regards as a "general principle for translating information into a probability assignment" (Jaynes 1989b: 211).

The Principle is meant to apply, not only to conditions in which we know nothing, but to conditions in which we have partial information about the probability distribution. For example, we might know the expectation value of some quantity. The Principle says that, given a partition $\{A_1, \ldots, A_n\}$ of possible events, if your prior information can be summed up via some set C of constraints, choose the probability distribution that, within those constraints, maximizes the Shannon entropy

$$S = -\sum_{i=1}^{n} Pr(A_i) \log(Pr(A_i)). \tag{3.1}$$

In the special case of no constraint, this is maximized by taking all probabilities equal, and so the Principle of Maximum Entropy reduces to the Principle of Indifference.

As a generalization of the Principle of Indifference, the Principle of Maximum Entropy inherits the same difficulties. The entropy of a probability distribution is defined relative to a partition, and so, the Principle of Maximum Entropy will yield incompatible probability distributions for different choices of partition. Applied to d'Alembert's game, the probability distribution that maximizes entropy relative to a partition consisting of the three possible outcomes of a game is the d'Alembert's folly distribution, assigning equal probabilities to each of the alternatives.

Even if we fix a partition as the "correct" one to use, there remains an element of arbitrariness to the Principle. It is simply ill-motivated. Consider this example from Jaynes (2003: 344). For a fair die, all faces are equally likely, and the expectation value of the number that lands face-up is

$$\frac{1}{6}(1 + 2 + 3 + 4 + 5 + 6) = \frac{7}{2} = 3.5$$

Suppose now, that you know that the die is biased in some way, but all you know is that the expectation value of the number that lands face-up is 4.5, instead of 3.5. How should you distribute your credence among each of the six faces?

We could, for example, assign Face 6 the highest probability, Face 1 the least, and have the remaining four faces have the same probability, between the probabilities of 1 and 6. This leaves considerable latitude, but one probability assignment that satisfies these conditions and yields the desired expectation value would be

$$(1/20, 1/8, 1/8, 1/8, 1/8, 9/20).$$

This is not the maximum entropy distribution. The distribution that maximizes entropy, subject to the constraint that the expectation value be 4.5, has the probabilities $(p_1, p_2, \ldots p_6)$ increase in a constant ratio that is equal to about 1.445. This gives a maximum entropy distribution that is (to three significant digits),

$$(0.0544, 0.0788, 0.114, 0.165, 0.240, 0.347).$$

This is a peculiar distribution. It's hard to see what credences about the mass distribution of the die might underly credences like this. The probability of Face 3 is closer to that of Face 4 than it is to Face 5, in spite of the fact that Faces 3 and 5 are adjacent, and Faces 3 and 4 are not.

A proponent of the Principle of Maximum Entropy might counter that the reason that we find this probability distribution peculiar is that we know something about the die that is not incorporated into the constraints. Nothing about the geometry of the die came into the calculation, and so the distribution does not take into account which faces are adjacent to each other. We also might have some suspicions about what sorts of biases can be produced by physically plausible causes, and we might somehow constrain plausible probability distributions according to these suspicions. However, it is implausible that such considerations can be encoded into a set of constraints that differentiates between the physically plausible distributions and those that do not. And, without a precisely delimited set of probability distributions to choose from, the Principle of Maximum Entropy fails to get off the ground.

But again, even in the cases in which we have a precisely definable set of constraints, the Principle of Maximum Entropy is simply an arbitrary stipulation.

3.1.6 Probabilities from symmetries

Perhaps physical symmetries will come to the rescue. In the case of a coin flip, we might note that the coin is, at least approximately, a squat cylinder, with a mass distribution that is rotationally symmetric about its vertical axis and, more importantly, symmetric about the plane that bisects its vertical axis.

But a symmetrical mass-distribution does not guarantee equal chances of *Heads* and *Tails*. Everything just said about the mass distribution remains true in the case of a magnetized coin in a magnetic field, where the complete situation will not have the same symmetry. In this case, there are relevant features that break the *Heads-Tails* symmetry.

But, of course, in any physical application, there will always be features that break the purported symmetry. It is too much to ask that the physical set-up be completely symmetric under interchange of *Heads* and *Tails*. (For one thing, in order to judge the outcome, the markings on the two sides of the coin must be different.) What is needed, for the application of an argument from symmetry, is approximate symmetry with respect to any features of the situation that are *relevant* to the outcome. Whether or not a given feature is relevant has to do with the dynamics of the process, and is not a merely geometrical feature.

Another asymmetry that will always be present in a coin toss lies in the initial conditions. The coin will start out with *Heads* facing up, or *Tails* facing up, or some other orientation. Is this a relevant asymmetry? To answer this question, no amount of intuition or purely *a priori* reasoning will help. We require either an investigation into the dynamics governing the coin toss, or empirical evidence concerning whether or not the coin has any disposition towards landing with the same side up, or, preferably, both. In the case of a simple coin toss, landing on a yielding surface without bouncing (so that the side that is up at the moment the coin first hits the ground is the side that is facing up when it comes to rest), both the physical analysis and the empirical evidence have been provided by Diaconis et al. (2007), who reach the conclusion: there is a small, but noticeable, bias towards landing with the same side up.

A judgment of equal chances is not an absence of judgment. But, armed with some idea of which factors are and aren't relevant to the chance of an event, considerations of symmetry can be a guide to reasoning about chances. If we supplement the abstract description of Bertrand's chord-choosing example with a specification of a physical set-up by which a chord may be chosen at random, then, as E. T. Jaynes (1973) has argued, considerations of plausibility of probability distributions may, indeed, yield a unique answer to the question posed.

3.1.7 The "empirical way"

Laplace, as we have seen, was not naïve about the need for a judgment about which cases are, in his terminology, "equally possible." Nor were the other seminal figures in the development of probability theory. There is a nice discussion of the issue in Jacob Bernoulli's *Ars Conjectandi*.

It was shown in the preceding chapters how, from the numbers of cases in which arguments for things can exist or not exist, indicate or not indicate, or also indicate the contrary, and from the forces of proving proportionate to them, the probabilities of things can be reduced to calculation and evaluated. From this it resulted that the only thing needed for correctly forming conjectures on any matter is to determine the numbers of these cases accurately and then to determine how much more easily some can happen than others. But here we come to a halt, because this can hardly ever be done. Indeed, it can hardly be done anywhere except in games of chance. The originators of these games took pains to make them equitable by arranging that the numbers of cases resulting in profit or loss be definite and known and that all the cases happen equally easily. But this no means takes place with most other effects that depend on the operation of nature or on human will.

... But what mortal, I ask, may determine, for example, the number of diseases, as if they were just as many cases, which may invade at any age the innumerable parts of the human body and which imply our death? And who can determine how much more easily one disease may kill than another—the plague compared to the dropsy, dropsy compared to fever? Who, then can form conjectures on the future state of life and death on this basis? Likewise, who will count the innumerable cases of the changes to which the air is subject every day and on this basis conjecture its future constitution after a month, not to say after a year? ... In these and similar situations, since they may depend on causes that are entirely hidden and that would forever mock our diligence by an innumerable variety of combinations, it would clearly be mad to want to learn anything in this way.

Nevertheless, another way is open to us by which we may obtain what is sought. What cannot be ascertained a priori, may at least be found out a posteriori from the results many times observed in similar situations.

(Bernoulli 2006: 326–7, translation of Bernoulli 1713: 223–4)

As we have seen in §2.4, in favourable circumstances chances can be estimated empirically. We do not have to reason *a priori* about which cases are equally possible, or which factors are relevant or irrelevant. If we suspect that varying some circumstance changes the chance of an event, we can put this suspicion to the test.

3.2 Why you are not a frequentist

It is not uncommon to find scientists making avowals to the effect that statements of probability are implicit references to long-run frequencies in some sequence of events. The purpose of this section is to convince you that, though frequencies have interesting and important links to chances, attempts to *define* chance in terms of frequency do not succeed. Taking "frequentist" to mean someone who adheres to a conception of probability on which the *meaning* of probability statements is to be cashed out in terms of frequency considerations, you are not a frequentist because, as I hope to persuade you, *there is no such conception*, despite valiant efforts by Venn, von Mises, and others to construct one, and that those who espouse frequentism are implicitly invoking a notion of probability other than frequency.

3.2.1 Frequentism and its problems

There are close connections between chances and frequencies. If there are n balls in an urn, m of which are black, then, if we draw one at random, with each ball having an equal chance of being the one drawn, then the chance that the ball that we draw will be black is m/n.

Chances are linked to frequencies in Bernoulli trials, also, via the laws of large numbers. If we flip a coin n times, and these are independent tosses, on each of which the chance of *Heads* is p, then, for large n, the chance is high that the relative frequency of heads will be close to p. If the sequence of tosses is prolonged indefinitely, then, with chance equal to one, the relative frequency of heads will converge to p. And, as we have seen, relative frequencies in a sequence of Bernoulli trials are effective evidence about chances.

All of these links between chance and frequency are noncontroversial; you can accept them and regard them as important without committing yourself to the existence of a frequency *interpretation* of probability. As I will be using the term, frequentism is the claim that we can *define* objective probabilities in terms of relative frequencies.[12] Versions of frequentism are

---

[12] A comment on terminology. At the present time there are two major approaches to statistics. The dominant approach, with roots in the work of Fisher and Neyman and Pearson, eschews talk of prior probabilities of hypotheses; the other, known as Bayesianism, permits

distinguished in two ways: the sequence of events considered may be actual or hypothetical, and they may be finite or infinite. In principle, therefore, there are four flavours of frequentism. In practice, however, they reduce to two. Frequentists of the actualist stripe usually do not assume that there are always actual infinite sequences of events of the right sort, and hypothetical sequences tend to be invoked in order to allow for infinite sequences. Thus we may restrict our attention to a finite frequentism that makes reference only to finite sets of actual events, and a hypothetical frequentism that makes reference to limiting frequencies in a hypothetical infinite sequence of events. Among notable defenders of frequentism are John Venn, who provided the first sustained defence of it, and Richard von Mises, its most influential proponent in the twentieth century.

The appeal of frequentism, at least in the finitist variety, is obvious. Frequencies are, undeniably, facts about the world, and so, if we're looking for objective probabilities, actual frequencies are an attractive candidate— provided, of course, that one *can* define probability in terms of actual frequencies.

The chief problem with frequentism, of either type, is that each of the links between chance and frequency mentioned indicates a relation between frequencies and chances that falls short of identity of concept. Crucially, each of them requires *a notion of chance distinct from relative frequency* to even state. Though considerable intellectual effort has gone into trying to construct a viable frequentism about chances, there's no way to get around this problem.

Take the first mentioned link between chance and frequency. We conclude that the chance of drawing a black ball is equal to the relative frequency of black balls in the urn, not merely on considerations of frequencies of balls in the urn, but on the assumption that each ball has an equal chance of being drawn. Without this, or some other assumption about the chances connected with the act of drawing that has the same effect (what's really needed is the assumption that, whatever else the chance of being drawn might depend on, it's independent of colour), there's no way to conclude

---

considerations of prior, and hence posterior, probabilities of the hypotheses under investigation. The former is often known as *frequentism*. This is an artifact of the fact, already mentioned in the previous chapter, that by the end of the nineteenth century, single-case chance had almost entirely dropped out of discussions of probability, and frequentism seemed to be the only option for objective probabilities. The methods of this school, however, are not wedded to a frequency conception of probability, and, in my opinion, make more sense if one thinks of the likelihoods it invokes as objective chances. Being a practitioner of so-called "frequentist" statistics does not commit you to frequentism as an interpretation of probability.

that the chance of drawing a black ball is equal to the relative frequency of balls in the urn.

Think, now, of Bernoulli trials, for example a sequence of independent coin flips, with *Heads* having the same chance of occurring on each trial. Then, by the Strong Law of Large Numbers, with chance equal to one, the limiting frequency of *Heads* in an infinite sequence of such trials, will be equal to the chance of *Heads* on each individual trial.

The value that the limiting frequency will almost surely have is equal to the single-case chance on each trial. Does this mean that we can *define* the chance of *Heads*, on any individual trial, as the limiting frequency of *Heads*? No. The problem is that a notion of chance, distinct from that of limiting frequency, occurs in the description of the set-up in two ways, and in the conclusion about limiting frequencies. One way in which the notion of chance occurs in the description of the set-up is that we assume that the outcomes of the trials are identically distributed, meaning that the chance of each of the possible outcomes is the same on each trial. The other is that we assume independence of outcomes of distinct trials. The conclusion about limiting frequencies is qualified by chance talk; a sequence on which there is a limiting frequency, let alone one on which there is a limiting frequency with the expected value, is not the only logical possibility; what licenses us to say that, if the series were to be continued indefinitely, the limiting frequencies of outcomes would approach values determined by the set-up, is the Strong Law of Large Numbers, which says that this would happen with *chance* equal to one.

These uses cannot be readily dispensed with.

The condition that we have a sequence of identically distributed, independent random variables is a sufficient condition for almost sure convergence, but not a necessary condition. It is easy to construct sequences for which there is a value that relative frequencies converge to with chance one, in which the chances are not equal on individual trials.

Consider, for example, the following set-up: *Drawing with Selective Replacement*. Consider an urn that initially contains 100 balls: 90 black balls, 6 white balls, and 4 red ones. Balls are drawn successively at random. Black balls are not replaced; the others are.

On an infinite sequence of such draws, with probability one, eventually all the black balls will have been drawn. From then on, with probability one, the relative frequencies of white and red will converge to 3/5 and 2/5, respectively. So, with probability one, the limiting relative frequency of draws of black, white, and red balls will be 0, 3/5, and 2/5, respectively. But these limiting relative frequencies are not their chances on the first draw, which are 9/10, 3/50, and 1/25.

It is for reasons like these that Venn made the move to hypothetical frequentism. In a human population, conditions may change from year to year, and the frequencies of events such as occurrence of diseases may exhibit secular variation. For this reason, Venn requires that we consider a series of events that is "regular throughout" (Venn 1866: 23). If there is no sequence of actual events that fits the bill, Venn enjoins us to substitute for the actual sequence an imaginary sequence that is the subject of our calculations. In a similar vein, in the motivating sections of his book, von Mises (1928, 1957) speaks of a "sequence of uniform events or processes" (1957: 12), though this drops out of his official definition.

What counts as a sequence that is "regular throughout" or "uniform"? Suppose that we changed the rules of our Drawing with Selective Replacement example so that black balls are replaced with other black balls. In this version, the number of balls of each colour remains the same on each draw. Presumably, each of the draws is relevantly of the same type. Why? Because the replacement of a black ball by another black ball does not change the chance of a black ball, or a ball of any other colour, being drawn.

Any two events will be similar in some respects and different in others. Unless we are given guidance about what is to count as a regular series, we have no clue how to construct one of Venn's hypothetical sequences. What, then, counts as a sequence of events of the requisite uniformity? One thing one could say (and this is what, I think, is really meant) is that two events count as the same type if and only if *they are similar in all respects relevant to their chances*. But, of course, this amounts to bringing the notion of *same chance* into our characterization of the sequences on the basis of which we are supposed to be defining chances, and hence (in my opinion, correctly) involves a rejection of the idea that *same chance* can be characterized in terms of frequencies.

3.2.2 The reference class problem

Another issue is the *reference class problem*, which affects both finite and hypothetical frequentism. Any event is a member of many different classes. Does the event have different probabilities, depending on which class it is thought of as a member of?

A serious frequentist, I think, has to bite the bullet and say that it does. This is von Mises' view.

This might seem to be unobjectionable. There's a way of thinking of it (which is not von Mises' way, but which has occurred to some) that seems to soften the bullet, but it's no help to a frequentist who thinks of frequentism as a way of making sense of objective probability, and wants to keep subjective elements out of it. That line of thought goes like this.

> Probability is relative to information. An insurance company who knows only that Ahmed is a male between the ages of twenty and thirty might assign one probability to the event of his being diagnosed with lung cancer within 5 years; his family and friends, who know that he is a chain smoker, might assign another probability. If all you know about someone is that the person is a member of a group, and you know that, of that group, a proportion p have a certain attribute, then you should assign probability p to the person having that attribute.

The reason that this doesn't help with frequentism, intended as an explication of objective probability, is that we've switched to an epistemic conception. This link between credence and frequency is operative within a purely epistemic conception. Suppose there is a collection of things, of which I know a proportion p has property X, and suppose that I assign equal credence to each distribution of the property X among the members of the group that is in accordance with the known proportion p. Then probabilistic coherence requires that my credence, for each member, that it has property X must be equal to p.

3.2.3 Some problems with finite frequentism

On finite frequentism, probabilities can only be rational numbers, and they can only be rational numbers whose denominators are equal to the number of elements in the class of events considered.

Consider a coin that is only ever to be tossed a finite number of times. If the number is even, according to finite frequentism it will be a fair coin only in the unlikely (in the ordinary sense of the word) event that the number of times that it lands heads is exactly equal to half of the total number of times it is tossed. If the coin is only tossed an odd number of times, it simply cannot be a fair coin. (The same reasoning, of course, applies if one widens the scope and considers all tosses of coins that are relevantly similar to the coin in question.)

Even if we are to bite the bullet and accept only rational probabilities, there is a problem that there might not be very many copies of the relevant type. It should be possible to construct peculiar gambling devices that are unique in the universe and which are only used once, or, perhaps, not at all. Should we not be able to ascribe non-trivial probabilities to outcomes of those devices?

There's another, serious mismatch between probability, as it appears in the textbooks, and finite frequentism. Suppose that we identified some finite class of events that we have decided provides the appropriate reference class for its members. For purposes of illustration, we may take it to be the class of all relevantly similar coin-tosses. Suppose there are N of these, in all history, and that, of these tosses, a fraction p of them come up *Heads*.

On the standard conception of probability, the one found in textbooks, it should be admissible to suppose these tosses to be identically, independently distributed. That is, it should be possible to suppose (whether or not it is true), that on each toss the probability of *Heads* is p, and that distinct tosses are probabilistically independent events.

But this is something that finite frequentism cannot do. If the tosses are i.i.d., then every possible frequency of *Heads* among the outcomes has a non-zero probability. This would be nonsense if one actually thought that what it *means* for the chance of *Heads* to be p is that, out of the entire sequence of N tosses, a fraction pN come out *Heads*.

The closest that a finite frequentist can come to a sequence of i.i.d. trials is to maintain that the probability of *Heads* on each trial is p, and that every sequence containing pN heads is equiprobable. On this probability assignment, the members of the sequence are treated symmetrically, but they are not independent. If we observe an initial segment consisting of a number m tosses, of which a number r yield *Heads*, then, conditional on this segment, the probability that the next toss will yield *Heads* is $(p - r/N)/(1 - m/N)$. If m is small compared to N, this is approximately p, and so we have approximate independence in the initial stages, but only approximate.

Finite frequentism runs into this problem because the link between probability of an outcome-type on an individual trial, and frequency of that outcome in the sequence, is not analytic. It is not the *meaning* of a statement about probability on an individual trial that the sequence of outcomes have any particular property. All possible sequences of outcomes may be assigned a non-zero probability, including those with properties that are not at all what one would expect.

3.2.4 Von Mises' frequentism

In Richard von Mises' *Probability, Statistics, and Truth* (first published in 1928), we find an attempt at a rigorous definition of probability in terms of hypothetical frequencies.

The motivating idea is that probability be nothing subjective, but, rather, be a physical property of events as real as any other physical property.

> The probability of a 6 is a physical property of a given die and is a property analogous to its mass, specific heat, or electrical resistance.
>
> (von Mises 1957: 14)

To ascertain the mass of a die, we take a sequence of measurements and average the results. Similarly, to ascertain the probability of a 6 upon rolling the die, we roll the die many times and consider the average rate of occurrence of 6.

So far, so good. However, von Mises then attempts to *define* properties such as the probability of 6 on rolling a die in terms of frequencies of results, with results as disastrous as operationalist attempts to define mass as the limiting result of a sequence of measurements.

For von Mises, an event considered in isolation cannot be ascribed any probability. In his slogan, "First the Collective—Then the Probability" (p. 18); it is only as a member of a collective (*Kollektiv*) that an event can be ascribed a probability at all. A collective is an infinite sequence of events satisfying two conditions. The first is the existence of limiting frequencies: the relative frequency of certain attributes of interest converges. The second is what von Mises calls the *Principle of Randomness*, or the *Principle of the Impossibility of a Gambling System*. The idea is that, if we pick out a subsequence according to any rule, where the selection may depend on earlier outcomes, the limiting frequencies in that subsequence will be the same as the limiting frequencies in the main sequence.

The definition of a collective is inspired by Bernoulli trials and the like. Because of the Strong Law of Large Numbers, with probability one the sequence of outcomes of Bernoulli trials will be one in which relative frequencies converge. Moreover, because of independence, for any given selection rule, the subsequence picked out by the rule is also a Bernoulli sequence, and hence, with probability one, the sequence of outcomes is such that limiting frequencies within that subsequence converge to the single-case chance.

There is a technical problem, which can be fixed. If the selection rule can be *any function*, then there are no sequences that satisfy von Mises' condition of randomness, except ones in which the limiting frequencies are all one or zero. For any sequence in which a given attribute occurs with a limiting frequency between zero and one, there will be subsequences with any desired limiting frequency.

However, for any *particular* selection rule, there are plenty of sequences that satisfy the condition, restricted to that rule. This can be extended to any countable collection of selection rules. One can fix von Mises' Principle of Randomness, while remaining in the spirit of the original suggestion, by restricting it to some countable set of selection rules, such as the set of rules that are effectively computable, as suggested by Church, or the set of rules, computable or not, that are definable in some formal system. This latter suggestion seems to be the one preferred by von Mises (pp. 92–3).

The fixed-up version, though, still fails to do justice to the *explicandum* that it is meant to explicate. For one thing, both conditions are expressed in terms of limits. As such, they are completely useless as guides to what we should expect of any finite initial sequence, because nothing is said about rates of convergence. You can take any infinite sequence satisfying von Mises' two conditions, tack on any finite initial sequence (for example, one with a length of $10^{10^{100}}$), and the new sequence will still satisfy von Mises' conditions, with the same limiting relative frequencies.

If we flip a coin 1,000 times, we expect the resulting relative frequency of heads to approximate the single-case chance of heads, and we adjust our credences about the bias of the coin accordingly, in light of the frequency data. But what this means is that we think that the observed relative frequency will *probably* be close to the single-case chance, and, for any degree of deviation of the observed relative frequency from the single-case chance, we can calculate the chance of such a deviation. But this requires a notion of probability distinct from limiting relative frequencies in an infinite sequence of coin flips.

A determined frequentist will respond by saying that the appropriate collective is an infinite sequence whose elements are repetitions of an experiment involving 1,000 coin flips, and the attributes in question are the number of *Heads* in each of these repetitions. In such a sequence, for each m from 0 to 1,000, the limiting frequency of the occurrence of m *Heads* will be equal to what we usually regard as the chance of getting m *Heads* in a single experiment involving 1,000 coin flips.

But, again, the claim about limiting frequencies in an infinite sequence of such experiments tells us nothing about what to expect in any finite sequence of such experiments. If I repeat the 1,000 coin flips 1,000 times, I would, for example, expect to see results containing no fewer than 480 Heads and no more than 520 Heads in about 800 of the trials. But this expectation is based on a calculation of probability, and, on von Mises' account, if I want to talk about the probability that N of the repetitions of the 1,000 coin flips will have a given relative frequency, I have to construct a new collective, a sequence, each of whose elements is a 1,000-fold repetition of 1,000 coin flips. And so on. To do justice to the *full probability distribution* on the space of possible outcomes of an infinite sequence of coin flips requires us to construct an infinite plethora of collectives.

Though inspired by considerations of independently identically distributed random variables, von Mises' defining conditions fail to capture either the idea that each member of the sequence is identically distributed, or that they are independent. To see this, let us return to the example of Drawing With Selective Replacement. With probability one, the limiting relative frequencies of black, white, and red will be 0, 3/5, and 2/5. Moreover, though the results of distinct draws are not independent, the correlations die off sufficiently quickly that, with probability one, the fixed-up version of von Mises' Principle of Randomness will be satisfied. With probability one, the sequence of outcomes of draws will satisfy von Mises' definition of a *Kollektiv*.

According to von Mises' definition, the probabilities of black, white, and red in a collective consisting of a typical result of this sequence of draws are 0, 3/5, and 2/5. Given that the very first draw is a member of this collective, this seems odd—we want to say that the probability of black on the first draw is 9/10. But remember: *First the collective, then the probability! Qua* member of the collective considered, the first draw has probability of black equal to the limiting relative frequency of black in the collective, that is, zero. If we want to say that the probability of black on the first draw is 9/10, we have to consider a different collective. Imagine that the Drawing With Selective Replacement scenario is repeated infinitely many times. We can form a collective consisting of first draws from each copy of the experiment. For a typical result of such an experiment, the relative frequency of black in *that* collective will be 9/10.

If you really take von Mises' definition to be adequate, you should be willing to bite the bullet. *Qua* member of one collective, the first draw inherits zero probability for black; *qua* member of another, it has the expected nine-tenths. There's no out-and-out-inconsistency here. But we've come a long way from the considerations that motivated the view. Von Mises wanted the probability of obtaining 6 on a roll of a given die to be a property of the die. On the theory he constructs, it is not: the probability will vary depending on which imaginary collective we take it to be part of. One could, similarly, want the probability of black upon drawing a ball from an urn to be determined by the physical properties of the urn, such as how many balls of each colour it contains, together with the manner of drawing on *that draw*, independently of whether we intend to draw with or without replacement on subsequent draws, and independently of which hypothetical collective we may be imagining it to be an element of. But, on von Mises' account, it is not.

3.2.5 Farewell to frequentism

As an *explicatum* of the notion of chance that motivates von Mises' account, his definition fails utterly. It fails to be objective; the probability of an event is not a property of an event but rather, a property of a hypothetical sequence of events that we construct in imagination. The same event can have different probabilities depending on which collective one imagines it to be a part of.

Though motivated by independently, identically distributed random variables, von Mises' definition fails to capture either the idea that the members of the sequence are identically distributed (a condition that is suggested by his introductory talk of sequences of uniform events, talk that drops out when he formulates his official definition), or the idea that the members of the sequence are independent, as independence is a sufficient condition for randomness in von Mises' sense, but not a necessary one.

Some readers may be tempted to exercise their cleverness in constructing a more complicated version of frequentism that evades these worries. My recommendation is that they not bother. The underlying motivation is deeply flawed, as it involves a conflation of the evidence we have for statements about chance—which is, indeed, usually frequency data—with the *meaning* of such statements.

3.2.6 Credits

There have been many critics of frequentism, and none of the arguments in this section is wholly original, though the presentation may differ from that of previous critics. Particularly influential for my own thinking about frequentism are Jeffrey (1992) and Hájek (1997; 2009).

4

What Could a "Natural Measure" Be?

4.1 The grand temptation

As we've already seen in Chapter 3, there's a temptation to think that the probability of an event can be *defined* as the ratio of the number of ways that the event can occur to the number of ways that the world could be. But this doesn't get off the ground without some judgment about which events are equiprobable events. In the case of a continuum, it requires a choice of measure. Most people, these days, would readily acknowledge that it is an illusion that probabilities can be defined, without further ado, as ratios of possibility-counts. It is an illusion reminiscent of the old illusion of Rationalism, that is, the idea that Pure Thought, without empirical input, can yield substantive knowledge about the world. Nevertheless, its influence lingers on.

If we allow the notorious dead horse to rest in peace, what could take its place? Here, again, we find, in the history of discussions of probability, clues as to the right track to take. Recall Bernoulli's remarks about cases that can happen with *equal facility*.[1] Which events those were could, according to Bernoulli, in some cases be judged by symmetry considerations, but, ultimately, were to be judged *a posteriori*.

There is, I claim, a way to make sense of the idea that certain events occur equally easily, others, more or less easily. This is not based on static consider-ations about the structure of the space of possibilities, but on considerations of *dynamics*. These considerations have to do with sensitivity to initial conditions, sensitivity that, for the right sort of dynamics, tends to wash out differences between probability distributions over initial conditions, in that very different probability distributions over initial microstates yield virtually the same probabilities for future values of certain macroscopic variables. There will (again, for the right sort of dynamics) be probability distributions over such variables that are stable under dynamical evolution

---

[1] See quotation in §3.1.7. For the history of locutions of this sort, which were common during the first century of the development of probability theory, see Hacking (1971).

Beyond Chance and Credence. Wayne C. Myrvold, Oxford University Press (2021).
© Wayne C. Myrvold.
DOI: 10.1093/oso/9780198865094.003.0004

and furthermore have the status of "attractor" distributions, meaning that other distributions tend to approach them. For a system of this sort that has been freely evolving long enough for this convergence to take effect (at least partially), we can use the attractor distribution to judge which eventualities occur with equal facility. It is considerations of this sort, I claim, rather than an appeal to a Principle of Indifference or any other considerations divorced from dynamics, that underwrite the use of standard probability distributions in statistical mechanics, and, indeed, in many situations in which we have well-defined probabilities.

Considerations of this sort will yield measures that are appropriate, and in some sense *natural*, for systems that have been evolving freely long enough for the requisite washing-out of disagreements among input distributions to have taken place. We have good reason to think that the standard measures evoked in equilibrium statistical mechanics are of this sort, a topic that we will take up in Chapter 8. There is no rationale, *none whatsoever*, for regarding them as privileged probability distributions for systems that have not been evolving freely long enough for some of the requisite washing-out to take place.

To convey a flavour of how this all works, it will be useful to consider a simple toy example I call the *parabola gadget*. Much of this chapter will be taken up with illustrating how the claims I wish to make about real systems are realized in this toy model.

4.2 The parabola gadget

Consider the device that I call the parabola gadget, depicted in Figure 4.1. It consists of a board, one meter square, on which is inscribed a diagonal. Also inscribed in the square is a parabola, which touches the two bottom corners of the square, and whose peak touches the top of the square in the middle. There is a ball that starts out on the diagonal, and moves according to the following rule. From the diagonal, it heads vertically (upwards or downwards, as need be) towards the parabola, until it reaches it. From the parabola, it heads horizontally (left or right) towards the diagonal, until it reaches it. The process then reiterates. In Figure 4.2(a) is shown one iteration of this process, and, in Figure 4.2(b), four iterations. Each iteration takes the same amount of time.

Suppose, now, you know that a parabola gadget has been running for some time, at least 50 iterations, and that you are asked which of the two alternatives you regard as more likely (see Figure 4.3):

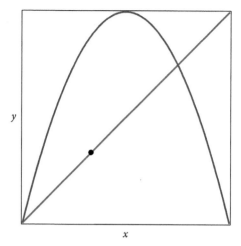

Figure 4.1. The parabola gadget

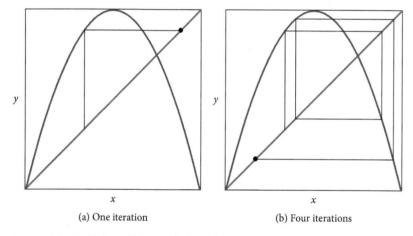

(a) One iteration   (b) Four iterations

Figure 4.2. Evolution of the parabola gadget

A. The ball is within 10 cm of the right-hand side.
B. The ball is within 10 cm, on either side, of the center.

I invite you, before proceeding, to give some consideration as to which you regard as more likely. Given a choice between a reward (something you like) if A is true, and the same reward if B is true, which choice would you make?

I imagine two agents, Alice and Bob, who make different choices. Bob reasons on the basis of a Principle of Indifference, as follows.

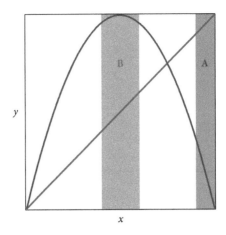

Figure 4.3. Alice and Bob's options

I know nothing about the initial conditions, and, even if I did know something, 50 iterations of the gadget would render that information useless, since small differences in initial conditions can lead to large differences in outcomes. In option B the payoff conditions span a range of positions twice as large as in option A, so option B is clearly preferable.

Alice, on the other hand, regards Bob's reasoning as seriously problematic, bordering on incoherence. Here's her thinking.

Though I am no fan of the Principle of Indifference, suppose I were to grant Bob the supposition that, at some iteration, say, the nth, I should regard intervals of equal length as equally likely. Then I can't say the same about the next iteration. It's clear from inspection of Figure 4.3 that all points that, at stage n, are in B, find themselves in A in the very next iteration. So, the ball's being in A at stage $n + 1$ must be at least as likely as its being in B at stage n.

Because of the shallowness of the slope of the parabola near its peak, points in some interval around the center get sent, in a single iteration of the machine, into a smaller interval near the right-hand side, which, on the next step, gets sent into a small interval near the left-hand side. There's a tendency for the ball to be more towards the edges than in the middle. On the basis of these considerations, A strikes me as more probable.

At the root of Alice's deliberations is the fact that, because of the dynamics of the machine, a probability distribution over the position of the ball at some time n uniquely determines a probability distribution over the position of the ball at time $n + 1$, as follows: the probability that, at time $n + 1$, the ball is in a set A, is equal to the probability that, at time n, it was at some point that gets mapped into a point in A by one iteration of the gadget's evolution. In this way, given a law of evolution of the state of some physical system, we can speak of the evolution of probability distributions over its state space.

Bob's favoured distribution is unstable. Applying it at some time n and also at time $n + 1$ is inconsistent with what Bob knows about the dynamics of the gadget. Suppose that he applies the uniform distribution to initial conditions. On this distribution, it is more likely that the ball will be in A 10 iterations down the line than that it will be in B: on a uniform measure over initial conditions, the set of states that put the ball into A after 10 iterations is larger than the set of states that put the ball into B after 10 iterations, by a factor of about 8 to 5. Both of these sets consist of a large number of small pieces, distributed over the length of the diagonal. For this reason, not only will a uniform distribution over initial conditions yield measures for these two sets that are roughly in the ratio 8/5, but the same holds for any probability distribution over initial conditions that doesn't vary "too quickly" with position, in a sense that can be made precise (see Appendix to this chapter).

It turns out that, though Bob's favoured distribution is unstable under evolution, there is another probability distribution that *is* stable.[2] Its density function is shown in Figure 4.4. As Alice has observed, it favors the regions near the edges. Call this invariant distribution μ.

If we consider various probability distributions over initial conditions and ask what they entail about probabilities for conditions at later times, we find that, for a wide class of distributions over initial conditions, the probabilities ascribed to states of affairs only a few iterations into the future closely approximate those of the invariant distribution μ. For any "sufficiently nice" distribution over initial conditions, this approximation gets closer, without limit, as one looks farther into the future.

For example, suppose that Bob adopts a uniform distribution over initial conditions. The density function for the probabilities this bestows on states of affairs 5 iterations into the future is shown in Figure 4.5(a). In Figures 4.5(b)

---

[2] There are others, e.g. the distribution that attributes probability one to the ball being *exactly* at the point of intersection of the parabola with the diagonal. But the invariant distribution we're concerned with is the only one that assigns probability zero to all sets of Lebesgue measure zero, and hence can be represented by a density function with respect to Lebesgue measure.

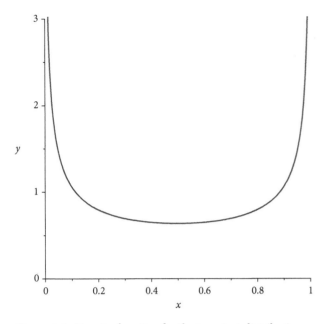

Figure 4.4. Density function for the invariant distribution μ

and (c) we see the effect of 5 iterations on other density functions for probabilities of initial conditions.

There's a theorem here: one can prove that, provided the probability distribution over initial conditions is represented by a density function that is not "too wiggly," then, for large n, what it says about the position of the ball n iterations into the future will be approximated by the invariant distribution μ, and, moreover, one can put bounds on how much it can depart from μ in terms of the wiggliness of the density function that yields probabilities over initial conditions. To be precise: one can show that, given a probability distribution over initial conditions that has a density f with respect to the invariant distribution μ, for any measurable set A, the probability, after n iterations, that the ball is in A, satisfies,

$$|Pr(x_n \in A) - \mu(A)| \leq \left(\frac{\mu(A)(1 - \mu(A))}{2^n} \right) V_T(f), \qquad (4.1)$$

where $V_T(f)$ is the total variation of f (see Appendix to this chapter for the definition of this, and for proof of this result).

Though we are not invoking a Principle of Indifference to mandate unique credences about initial conditions, it is reasonable to expect that, even if

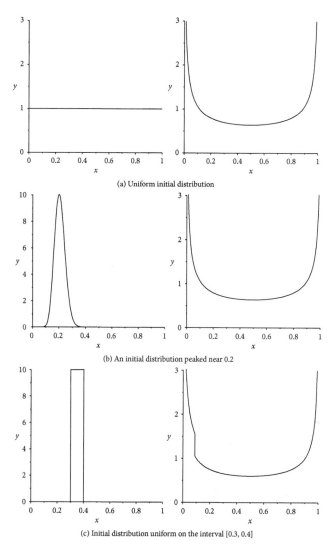

(a) Uniform initial distribution

(b) An initial distribution peaked near 0.2

(c) Initial distribution uniform on the interval [0.3, 0.4]

Figure 4.5. The effect of five iterations of the parabola gadget on various input distributions

Alice and Bob know little about the process that sets the initial conditions, they know enough about the process that they don't expect it to reliably target minute regions of the diagonal. This will be reflected in a credence function that is represented by a continuous density function that doesn't vary excessively quickly as one moves across the diagonal. How quickly is excessively quickly depends on what a reasonable person could believe about the sorts of processes that could set initial conditions at the time the gadgets are set running.

Only a little uncertainty is needed to get the method off the ground. For an illustration, suppose we change the example a bit, and allow Alice and Bob to measure the initial position of the ball. Let's be generous, and allow them a high-precision measuring instrument, a device with a normal distribution of measurement errors with a standard deviation of one nanometer.

Assume that, upon receiving a read-out from this device, the information supercedes any prior credences they might have about the position of the ball, and that their new credences about its position can be approximated by the error distribution of the measuring device. That is, near the observed value, their credences can be approximated by a normal distribution with standard deviation one nanometer. Any such distribution has a density function, with respect to μ, that has total variation satisfying

$$V_T \leq \sqrt{\frac{\pi}{8}} \, 10^9. \tag{4.2}$$

If we apply the convergence result, Corollary 4.1.2 of the Appendix, we find that, given an input distribution with this total variation, the probability that the ball is in A after 32 iterations (or more) must be greater than the probability that it is in B, no matter what the result of the measurement is.

If we take it that, in the absence of a source of information about the position of the ball more precise than the measuring device considered, reasonable credences about the position should not have greater total variation than credences informed by the result of a measurement by such a device, then all reasonable credences about the initial position of the ball will have total variation satisfying (4.2). The idea is that you should not pretend to more precise knowledge than you could possibly have.

For a gadget that has been running sufficiently long, probabilities about the position of the ball are largely independent of what one takes to be an appropriate measure over initial conditions, and one can use μ to calculate these probabilities. In this sense the dynamics of the gadget pick out μ as a "natural measure" for gadgets that have been running for a while. This is not because μ is singled out as a natural measure over initial conditions; we might be rather vague about what to think about initial conditions. It is, rather, because judgments about probabilities about the state of a gadget that has been running for a while are largely (though not entirely) independent of probabilities of initial conditions. On a wide variety of ways of measuring sizes of sets of input states, including, but not restricted to, a measure uniform in distance along the diagonal, there is a larger measure of initial conditions that put the ball 10 cm from the right edge than of those that put it in an interval of 20 cm centered on the midpoint of the gadget.

To adopt Bernoulli's phraseology: things like the ball being, after many iterations of the evolution, in a 10 cm interval near the center do *not* occur with equal facility as the ball being in an interval of the same length near an edge. If we were try to place the ball so precisely that, after 32 iterations, it is more likely to be in B than in A, a hand with a slight shake on the order of a nanometer will thwart our purpose.

4.3 Sensitive dependence as a source of predictability

Suppose I have 1,000 gadgets, each of which has been running independently for some time. Consider "macrostates" of this system: we partition the width of the box into 20 intervals of 5 cm, and specify, for each interval, how many of the balls, out of the 1,000 gadgets, lie within that interval.

Suppose, now, I ask you to make a prediction about the macrostate. You get to choose between two propositions about the macrostate.

A'. More of the balls will be in region A than in region B.
B'. More of the balls will be in region B than in region A.

A naïve application of the Principle of Indifference of the sort favored by Bob, on which, for each gadget, equal intervals of the diagonal are equally likely, yields a measure on which the set of states that make B' true is vastly larger than the set that makes A' true. But, as, we have seen, we should not use such a measure for gadgets that have been running for a while.

If you grant the reasoning of the previous section, then you should, for each of the gadgets, regard A as about 8/5 times more likely than B. More-over, the evolution of the gadgets will tend to erase correlations between initial states of distinct gadgets, so you should take probabilities regarding one gadget to be independent of probabilities regarding any other gadget. Given probabilities satisfying these conditions, it is highly probable that the number of gadgets with balls in A will be close to 8/5 times the number of gadgets with balls in B (this is, of course, an application of the Weak Law of

Large Numbers), and so you can be highly confident that more of the gadgets will have their balls in A than in B. You should regard A' as overwhelmingly more probable than B'.

Let M be the probability measure over the set of 1,000 gadgets on which each gadget is independently endowed with the invariant measure μ. On measure M, the set of states in A' is vastly larger than the set of states in B'. Moreover, any measure over initial conditions that is not too crazy will tend to agree, to a close approximation, with M about probabilities of states of affairs after a few iterations. This means that on any measure over initial conditions of the collection of gadgets that isn't too crazy, the set of initial conditions that lead, after 10 or so iterations, to states in A' is vastly larger than the set of states that lead to states in B', and, for any such measure, we can use M to compute approximately how much larger. Again: this isn't because M is favoured as a "natural measure" over initial conditions; this is a conclusion that is largely (though, of course, not entirely) independent of choice of measure over initial conditions.

In this way, we get predictability, with near certainty, of certain aspects of the state of a system consisting of a large number of parabola gadgets that have been running for an appreciable time, not *in spite of* sensitive dependence on initial conditions, but *because of it*.

4.4 Invertibility

So far we have talked only about probabilities over initial conditions—that is, over the state of the gadget when it is set running—and their implications for probabilities for future states. Some readers will be wondering about how things might go in the other temporal direction.

Given what has been said before, readers may be forgiven for thinking that, since each position on the diagonal (except the extreme point on the right, which can be reached only from the center) can be the result of two previous positions, the evolution of the parabola gadget is not invertible. However, there is a detail that I have so far not mentioned. In addition to the moving ball, there is a pointer that shuttles back and forth along the left edge of the gadget. Call the distance (in meters) of the pointer from the bottom edge, z. Its value changes as follows. If, at time n, the ball is to the left of center (or exactly on center), in the next iteration the value of z is halved; that is, the new position of the pointer is a distance $z/2$ from the bottom. If, at time n, the ball is to the right of center, the position of the pointer is a distance $z/2$ from the top. Thus, the position of the pointer at time $n + 1$ carries information about the position of the ball at time n. If, at time $n + 1$, the pointer is in the bottom half, then the ball, at time n, was in the left half, and, if at time $n + 1$, the pointer is in the top half or at the center, the ball, at time n, was in the right half or at the center. Each position of the pointer at time $n + 1$ can be reached from one and only one position at time n. Therefore, the state of the gadget at time $n + 1$ uniquely determines its state at time n. Since the state of the gadget at any stage of its running uniquely determines its state at all earlier and later stages, a probability distribution over the state (x, z) at some time uniquely determines probabilities for all earlier and later states, so long as the gadget runs undisturbed during the interim.

Forward evolution of the gadget leads to a situation in which the probability distribution for x is closely approximated by μ, on which the ball is equally likely to be on either side of the diagonal. Thus, after sufficiently many iterations, any information about the past of the gadget gets buried very deeply in the fine details of the distribution of z, and, for any interval $[a, b]$, the probability that z is in that interval approaches the length of the interval. The uniform distribution over z is an attractor distribution.

Let ρ be the probability distribution on which x is distributed according to μ and z is uniformly distributed, independently of x. This is an attractor distribution over the state space of the gadget. It can be proven (see Appendix to this chapter) that any probability distribution that can be represented by a density function that is sufficiently nice converges towards this invariant distribution, in the sense that, for any measurable subset A of the unit square, the probability that the state (x, z) at time n will be in A approaches $\rho(A)$, as n becomes large.[3]

Evolution of the gadget does not, however, tend to smooth out probability density functions for z. Suppose, for example, that you are sure that initially the ball is to the left of center. Then, one step in, z will be in the lower half of its range. After two steps, z must be in either $[0, 1/4]$ or $[3/4, 1]$, and after 3, in $[0, 1/8]$ or $[3/8, 5/8]$ or $[7/8, 1]$. After a large number of steps, the support of the probability distribution for z will be highly fragmented,

---

[3] Resist the temptation to reverse the order of the quantifiers! For any sufficiently nice input distribution, for any measurable set A and any $\varepsilon > 0$, there exists N such that, for all $n > N$, the probability that the state at time n is in A is within ε of $\rho(A)$. However, if the input distribution differs from ρ, and there is a set B such that the probability that the initial state is in B differs from $\rho(B)$ by some amount δ, then, for any n, no matter how large, there will be some set B_n such that $\rho(B_n) = \rho(B)$ and the probability that the state at time n is in B_n differs from $\rho(B_n)$ by δ. For more on this, see §8.2.1.

in such a way that any interval that is not too small will be half-covered by this support. This is the way that the distribution of z converges towards a uniform distribution. As proven in the Appendix, it tends to go at a slower rate than the convergence of x towards its equilibrium distribution.

Since the dynamics are invertible, we can also back-evolve probability distributions. In the reverse direction we also get convergence towards the equilibrium distribution. The condition for convergence in the backwards direction is that the density function for z not be too wiggly.

Whereas forward evolution tends to turn probability density functions over x into ones that closely resemble the density function for μ and tends to complicate density functions for z, backwards evolution tends to smooth out density functions for z and complicate density functions for x. For example, consider a probability distribution that yields certainty, at some time $t + n$, that the ball is to the left, and ask what probabilities over states of affairs at time t could lead to such a thing. Because of convergence towards the equilibrium distribution in the forward direction, we know that it will have to have a very wiggly density function for x. Figure 4.6, shows, by way of

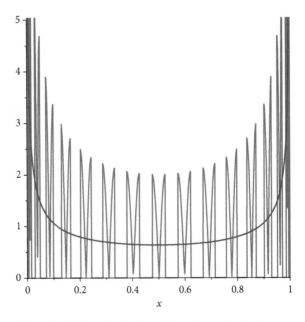

Figure 4.6. Five-step backwards evolute of a density function uniform over left half of the diagonal

example, the density function that results from back-evolving by 5 steps a distribution that is uniform in z and, in x, uniform over the left half of the diagonal. It is very wiggly, but, on a coarse-grained level, approximates the invariant distribution μ, in that intervals that are not too small are accorded roughly the same probability by this distributions as by μ, whose density function is also shown in Figure 4.6, for comparison.

To sum up:

- The measure ρ is invariant under evolution.
- It is also an attractor distribution in both forward and backwards directions. Given a probability measure over the state of the system at some time t:
  - As long as the density function for the value of x at time t is not too wiggly, the probability that the state of the system being in a set A at later time $t + n$ is approximately equal to $\rho(A)$ for large n.
  - As long as the density function for the value of z at time t is not too wiggly, the probability that the state of the system being in a set A at earlier time $t - n$ is approximately equal to $\rho(A)$ for large n.

At first glance, convergence towards an equilibrium distribution in both forward and reverse directions might seem paradoxical, perhaps contradictory. Suppose, for example, we have, at time t, a distribution that is very different from the equilibrium distribution. Evolve it forward n steps, far enough into the future that the evolved distribution is a good approximation to the equilibrium distribution. If we back-evolve this n steps, will we not get something that approximates the equilibrium distribution, instead of recovering the distribution we started with?

The attentive reader will already know how to resolve this apparent paradox. The forward-evolved distribution will incorporate detailed information about the history of the system, and this will be represented by a very complicated distribution for z. The density function for z yielded by this forward-evolved distribution will be sufficiently complicated that it will take more than n steps to back-evolve it into anything like a uniform distribution.

An example might help make this vivid. Suppose that at time t, you observe that the ball is in the left-hand side of the box. Let it evolve 100 steps forward. Then your knowledge of the past of the system entails something about its current state; in particular, you know that the value of z is in the finely disaggregated sets of states that record the previous location of the ball. If your ability to measure the value of z is limited to an instrument with a resolution of one nanometer, your knowledge of the past affords more detail about the value of z than could be afforded by measurement. In such a case, both of the following may be true: (1) Any reasonable credence function based on nothing other than a direct observation of the current state of the gadget, when evolved backwards 100 steps, yields a close approximation to the attractor distribution ρ, and (2) Your credence is reasonable, and, when back-evolved, yields certainty (or close to it) that the ball was in the left half, 100 steps ago.

The mathematical facts remain pretty much the same with the temporal orientations reversed. Suppose that, somehow, perhaps via an oracle, you knew that, 100 steps to the future, the ball will be in the left half. This would tell you something about the current state of the gadget that could not have been obtained by direct measurement, namely, that it is within the fragmented subset of states that evolve, in 100 steps, into the left side of the gadget. This would be represented by a density function over x at the current time that is so wiggly that forward evolution by 100 steps is not sufficient to bring it into an approximation of the invariant distribution μ.

If you have knowledge of the past of the system, you may have very detailed knowledge of its state, but this knowledge is of no avail in making predictions. This is illustrated in Figure 4.7. In the interest of graphics resolution, we illustrate the situation for 5 rather than 100 steps. Figure 4.7(a) depicts what you know about the state of the system if you know that the ball was in the left-hand side of the box, 5 iterations ago; you know that the current state is within the shaded area. Figure 4.7(b) depicts what you would need to know about the state of the system to be sure that it will be in the left-hand side of the box, after 5 more iterations. You need to be certain that the current state is within the shaded area. There is no correlation between what you know and what you need to know to make a reliable prediction; on the measure ρ, the measure of the intersection of these two sets is just the product of their respective measures.

To sum up: if you have knowledge of the state of the gadget n steps to the past of a time t, of the sort obtained by some other means than taking an observed state at t and back-evolving it, this knowledge gets encoded in a probability distribution over the state at time t via a complicated distribution for z, a distribution that is sufficiently complicated that back-evolving it n steps does not yield a close approximation to the attractor distribution ρ. Similarly, if you have knowledge of the future of a time t, of the sort obtained by some means other than taking an observed state at t and forward-evolving it, then this would be encoded by a probability distribution over

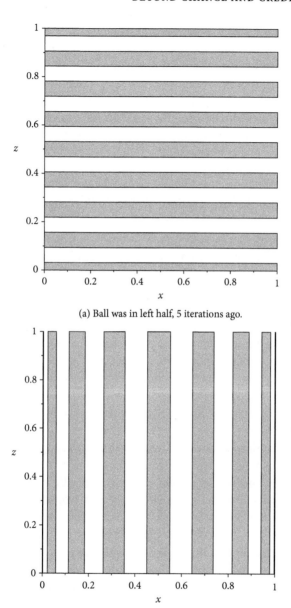

(a) Ball was in left half, 5 iterations ago.

(b) Ball will be in left half, 5 iterations to the future.

Figure 4.7. The difference between knowledge of the past and of the future

the state at time t with a complicated distribution for x, a distribution that is sufficiently complicated that forward-evolving it n steps does not yield a close approximation to the attractor distribution ρ.

Our convergence results themselves do not care about temporal direction. But in application, there will be a temporal asymmetry, due to the fact that when it comes to predictions, we don't have access to oracles, and we can do no better than take our knowledge of the current state and use our knowledge of the dynamics to forward-evolve it.

4.5 Invariant distributions as surrogates

Take a probability distribution over conditions at some time t, with probabilities for x that are represented by a none-too-wiggly density function. Let n be a number of steps that is sufficiently large that probabilities for conditions n steps to the future of t are closely approximated by the equilibrium distribution. If the distribution for conditions at time t is very different from the equilibrium distribution, the distribution over z at time $t + n$ will be very complicated in its details, though it will approximate the equilibrium distribution at a coarse-grained level. These fine details will, however, be largely irrelevant for calculating probabilities over conditions at times to the future of time $t + n$, and, for the purposes of such calculations, we can replace the complicated distribution over conditions at $t + n$ by the equilibrium distribution.

To the extent that, at time $t + n$, details of the system's past have become irrelevant for its future behaviour, we can discard information about its past and use the equilibrium distribution as a surrogate for the more complicated one that encodes information about the past. It would, of course, be a mistake, when making retrodictions, to discard information about the past and to back-evolve a smoothed-out distribution.

For beings such as ourselves, who have access to records of the past but, when it comes to predicting the future, typically can do no better than to take the current conditions and forward-evolve them, there will be an asymmetry in the invocation of convergence results. We can use a simple distribution as a surrogate for a more complicated one when it comes to predictions, insofar as the information discarded is irrelevant for future predictions, but, when it comes to retrodictions, this would be nothing short of madness, as we would be discarding relevant information.

4.6 Introducing uniform distributions

The dynamics of the parabola gadget pick out a measure that is appropriate to use when making predictions concerning the future of a gadget that has been evolving freely for some time. Though uniform in z, it is not uniform in x.

Of course, whether or not a distribution is uniform depends on the parameters used to characterize the state. The state of a system may equally well be represented by a different choice of variables. If one is enamored of uniform distributions, one can indicate positions along the diagonal via a new variable, u, defined by

$$x = \sin^2(\pi u/2). \tag{4.3}$$

This function is graphed in Figure 4.8. On the invariant distribution μ, u is uniformly distributed: equally sized intervals of u have equal probability. As a result, when working with probability distributions over the state of the gadget, it can be more convenient to work with (u, z) rather than (x, z).

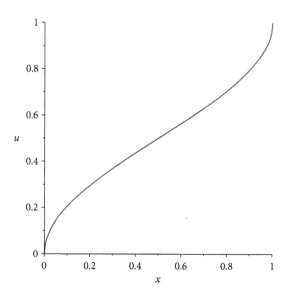

Figure 4.8. The function $u(x)$

4.7 Status of the input distributions

I have claimed, and demonstrate in the Appendix, that a wide range of probability distributions over initial conditions of the parabola gadget yield convergent probabilities for positions of the ball at later times. The dynamics map probability distributions at one time to distributions at other times, but this map requires probabilities as inputs. With no probabilities in we get no probabilities out. What, then, is the status of the input distributions we have invoked?

Following a suggestion of Savage (1973), we have been talking as if these are credences, or subjective degrees of belief. The distributions that result from evolving credences about states of affairs at time t_0, via the actual dynamics of a physical system (dynamics that might be unknown, or imperfectly known, to an agent, who might not be able to do the calculation even if the dynamics are known) are things that partake of both epistemic and physical considerations. There is an epistemic aspect, as some uncertainty about the state of the system is required. But they need not be the actual credences of any agent, because, as mentioned, an agent might not know what the result is of evolving her credences via the actual dynamics. However, in the sorts of cases we're interested in, this evolution will tend to minimize individual differences between agents' credences, and the values to which probabilities converge are determined by the dynamics of the system. These sorts of probabilities have been called *almost objective probabilities*. To emphasize that they have both epistemic and physical aspects, I have elsewhere called them *epistemic chances* (Myrvold 2012a).

All we need is some uncertainty, perhaps of a vague degree, in the agents' knowledge of initial conditions, plus some (perhaps vague) sense of a range of credence functions that are reasonable, in light of that uncertainty, and we're off and running. This need not be a purely subjective matter; judgments about what sorts of credences about initial conditions are reasonable are based on judgments about what sorts of processes there are that could lead up to those conditions.

For some systems, a classical treatment will be inadequate, and our discussion will have to be cast in terms of quantum mechanics. Such a treatment will run in much the same way. We never know for certain the precise quantum state of system. What we will want from the quantum evolution will be that it take a wide variety of initial quantum states into ones whose restriction to macrovariables is roughly the same. Thus, even if it can be argued that all probabilities in statistical mechanics, even classical statistical mechanics, have their source in quantum mechanics, convergence results of the sort we've been discussing will play a central role.

On could also take seriously (as I think we should) the thought that the fundamental laws of physics are not deterministic, but stochastic. You might imagine, for example, that the dynamics given for the parabola gadget are not exact—perhaps, instead of dictating the position of the ball on the next iteration, the dynamics yields a probability distribution for the position, such that the position of the ball on iteration $n + 1$, given its position on iteration n, may differ from the one required by the dynamics we have been discussing by a slight deviation, a sort of Epicurean swerve, whose exact value is a matter of chance. For a theory like that, the *dynamics alone* will place limitations on how much one could know about the state of a system that has been evolving for a while, because the dynamics alone will produce a range of possible states from one and the same initial state.

The dynamics of such a theory will produce some uncertainty. But, if we are going to get convergence towards certain probability distributions over the state of the system, we will still need convergence of probability results like the ones discussed in this chapter. More on this, later, in Chapter 9, which deals with the quantum realm, where we have good physical reason to take seriously the idea that the laws of physics are indeterministic.

4.8 Partial equilibration

Consider a probability distribution that is initially concentrated on some sub-interval of the diagonal, such as the one whose density function is depicted in Figure 4.9(a); this one is confined to the interval $[0.3, 0.4]$, and is uniform, in position along the diagonal, on that interval. It takes several iterations of the gadget to spread this distribution over the full width of the diagonal. But something interesting happens in the meantime. Take a look at Figure 4.9(b). There we see the one-step evolute of the distribution of 4.9(a). It is confined to the interval $[0.84, 0.96]$, but, on that interval, it closely approximates the restriction of μ to that interval.

Call this phenomenon *partial equilibration*. The probability distribution is nothing like the equilibrium distribution, as the position is confined to a subinterval. But, subject to that constraint, it comes near to being as much like the invariant distribution as it could be, while satisfying that constraint. We can, without significant loss, replace the probability distribution for x by one that is the restriction of the invariant distribution to that interval.

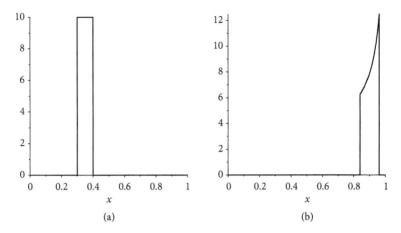

Figure 4.9. Local equilibration

As the variable z is irrelevant to the evolution of x, we can use a uniform distribution in z. The resulting distribution over the complete state space of the system is one that is as close to the equilibrium distribution as can be, given the restriction that the value of x is confined to a certain sub-interval. In some ways, this is very different from a distribution that represents what we actually know about the state of the system. Since we know that the ball was, in the previous step, on the right half of the diagonal, this has implications for the value of z that are not reflected in this smoothed-out distribution. These differences are, however, irrelevant for probabilities concerning future values of x, and hence may be disregarded when it comes to prediction.

4.9 Conclusion

If we are to regard certain measures as "natural" ones to use, for a given physical system, no Principle of Indifference can come to our aid. Dynamics can, however. For certain sorts of systems, with the right sorts of dynamics, there is a measure that is picked out by the dynamics as appropriate, provided that the system has been evolving freely long enough for at least partial equilibration to take place. In such a circumstance, the natural measure can be used as a surrogate for a measure that represents what we actually know about the system, as long as it is used only for prediction of future states. There is no rationale for taking a measure of this sort as privileged when dealing with systems that have not undergone the requisite partial equilibration, or for using it for retrodiction.

4.10 Appendix

4.10.1 Evolution of density functions: the
Frobenius–Perron equation

Suppose we have a map $T : \mathbb{R} \to \mathbb{R}$, and a sequence of probability distributions on $\langle \mathbb{R}, \mathcal{B} \rangle$, related by

$$P_{n+1}(A) = P_n(T^{-1}(A)). \tag{4.4}$$

Suppose that these distributions have density functions f_n with respect to Lebesgue measure. How are these related?

We won't assume that the map is invertible. For any x, let $\{T_i^{-1}(x)\}$ be the functions that produce all the inverse points of x. That is, $T(T_i^{-1}(x)) = x$, and all distinct points that are mapped to x are the values of exactly one $T_i(x)$. Then the densities are related by,

$$f_{n+1}(x) = \sum_i f_n(T_i^{-1}(x)) \left| \frac{d}{dx} T_i^{-1}(x) \right|. \tag{4.5}$$

This is known as the *Frobenius–Perron* equation.

For the parabola gadget, the mapping of the diagonal to itself, on one iteration of the gadget's evolution, is

$$x_{n+1} = T(x_n) = 1 - 4(x_n - 1/2)^2 = 4x_n(1 - x_n). \tag{4.6}$$

This has inverse functions

$$\begin{aligned} T_1^{-1}(x) &= (1 - \sqrt{1 - x})/2; \\ T_2^{-1}(x) &= (1 + \sqrt{1 - x})/2. \end{aligned} \tag{4.7}$$

This gives us

$$f_{n+1}(x) = \frac{1}{4\sqrt{1 - x}} \left(f_n((1 - \sqrt{1 - x})/2) + f_n((1 + \sqrt{1 - x})/2) \right), \tag{4.8}$$

which is what was used to compute the evolved density functions graphed in this chapter.

4.10.2   Proof of the convergence result

We will now make precise and prove the convergence results alluded to in this chapter. For related results, see Engel (1992) (in particular, Theorem 3.9a and §3.2.6).

Let (x_n, z_n) be the state of the gadget at stage n of its evolution. The dynamics specified in the main text gives

$$x_{n+1} = T(x_n) = 1 - 4(x_n - 1/2)^2 = 4x_n(1 - x_n). \tag{4.9}$$

$$z_{n+1} = \begin{cases} z_n/2, & \text{if } x_n \leq 1/2; \\ 1 - z_n/2, & \text{if } x_n > 1/2. \end{cases} \tag{4.10}$$

Readers familiar with the literature on chaos theory will already have recognized (4.9) as the logistic map.

A probability distribution for x_0 determines distributions for each x_n, $n > 0$. Suppose that we have a probability distribution for x_0 that has density f with respect to the invariant measure μ. That is, for any measurable subset A of the unit interval,

$$Pr(x_0 \in A) = \int_A f(x) \, d\mu(x). \tag{4.11}$$

If g is the corresponding density with respect to λ_x, that is, Lebesgue measure induced by the parameter x, the two are related by

$$f(x) = \pi\sqrt{x(1 - x)} \, g(x). \tag{4.12}$$

Thus, the invariant distribution, which has flat density with respect to itself, has density, with respect to λ_x,

$$g_\mu(x) = \frac{1}{\pi\sqrt{x(1 - x)}}, \tag{4.13}$$

which is the function we have seen graphed in Figure 4.4.

However, it is the density f, the density with respect to the invariant measure μ, that will be of interest to us, as it is in terms of this density that we obtains bounds relevant to rates of convergence.

It will be useful to make a change of variables. Define the variable u by

$$x = \sin^2\left(\frac{\pi u}{2}\right). \tag{4.14}$$

This variable is useful because the invariant distribution μ is uniform in u. That is, for any interval $[a, b]$ within the unit interval,

$$\mu(\{u : u \in [a, b]\}) = b - a. \tag{4.15}$$

The evolution (4.9) induces a corresponding map for u:

$$\sin^2\left(\frac{\pi u_{n+1}}{2}\right) = 4\sin^2\left(\frac{\pi u_n}{2}\right)\cos^2\left(\frac{\pi u_n}{2}\right) = \sin^2(\pi u_n). \tag{4.16}$$

This gives,

$$u_{n+1} = 1 - 2|u_n - 1/2| = \begin{cases} 2u_n, & u_n \le 1/2; \\ 2(1 - u_n), & u_n > 1/2. \end{cases} \tag{4.17}$$

This is the tent map. It's easy to work with because it's piecewise linear.

The invariant measure ρ is uniform in u and z. It can be shown that the evolution is *strong mixing* with respect to ρ: that is, for any measurable sets A, B,

$$\rho(A_n \cap B) \to \rho(A)\rho(B) \text{ as } n \to \infty, \tag{4.18}$$

where A_n is the result of applying n iterations of the evolution to A. Hopf's proof (1934, §8) that the baker's map is strong mixing applies equally well to the evolution of the gadget. From this it follows that any distribution over initial conditions that has a density with respect to the invariant measure converges towards this measure. That proof doesn't, without further ado, provide information about bounds on rates of convergence, which we now investigate.

Iteration of the tent map n times produces 2^{n-1} copies of the tent, each supported on an interval of length $2^{-(n-1)}$. Each of these intervals consists of two sub-intervals of length 2^{-n} that are mapped linearly onto the unit interval. Let Δ_i, for $i = 1, \ldots, 2^n$, be the sub-interval $[(i-1)/2^n, i/2^n)$.

Now, let us consider some measurable subset B of the unit interval, with measure $\mu(B)$, and consider its inverse image under n-fold iteration of the tent map; that is, consider the set A that gets mapped into B. Each of the sub-intervals Δ_i contains a subset $A_i = A \cap \Delta_i$ of measure $\mu(B)/2^n$. The probability that u_n is in B is equal to the probability that u_0 is in A, which is the sum of the probabilities of $u_0 \in A_i$ over all the sub-intervals A_i.

Expressed in terms of our original variable x: n-fold iteration of the logistic map partitions the unit interval into 2^{n-1} intervals of equal μ-measure $2^{-(n-1)}$, each of which contains two sub-intervals of measure 2^{-n} that get mapped onto the unit interval. For any measurable subset B, with inverse image A, each of these sub-intervals contains a subset of A of measure $\mu(B)/2^n$.

We want to investigate the probability that a distribution with a given density function f ascribes to A. We will make use of the following theorem.

Theorem 4.1 *Let x be a random variable that has density f with respect to some measure μ. Let A be a measurable set with the property that the range of x can be partitioned into subsets Δ_i such that, for each i,*

$$\mu(\Delta_i \cap A) = \mu(A)\,\mu(\Delta_i).$$

Let f_i^+, f_i^-, be the essential supremum and essential infimum, respectively, of f on Δ_i. Then

$$|Pr(x \in A) - \mu(A)| \leq \mu(A)\,\mu(\bar{A}) \sum_i \mu(\Delta_i)(f_i^+ - f_i^-),$$

where $\bar{A}$ is the complement of A.

From this follows a corollary in terms of the *total variation* of the density function f. Alpine hikers will find the concept of total variation intuitive. Imagine walking along the graph of the function from left to right. The total variation is the total vertical ascent plus the total vertical descent you have to do. The official definition (which is found in many textbooks of real analysis; see e.g. Kolmogorov and Fomin 1975: §9.32) is as follows.

Definition. Consider a function $f : [a, b] \to \mathbb{R}$, defined on some interval $[a, b]$ of the real line. Take any finite increasing set of numbers $a = x_0 < x_1 < \ldots < x_n = b$, and consider the quantity

$$\sum_{k=1}^{n} |f(x_k) - f(x_{k-1})|.$$

The *total variation* of f, $V_T(f)$, is defined to be the essential supremum of this quantity, over all choices of $x_0, \ldots, x_n$.

Obviously, a constant function has total variation zero. If a function has finite total variation, it is said to be of *bounded variation*. The density function of the invariant measure μ with respect to λ_x, given in equation (4.13), has unbounded total variation, but its density function with respect to itself is flat and has total variation zero.

A function of bounded variation on $[a, b]$ has a finite derivative at almost all points in $[a, b]$ (Kolmogorov and Fomin 1975: 331). If f is a continuous function of bounded variation, then

$$V_T(f) = \int_a^b |f'(x)| \, dx.$$

If f is piecewise continuous, we add to this the sum of the absolute values of the jumps that f makes at each of its points of discontinuity.

From the above theorem the following corollary is immediate.

Corollary 4.1.1 *Under the conditions of Theorem 4.1, if f has finite total variation $V_T(f)$, and if there exists δ such that $\mu(\Delta_i) \leq \delta$ for all i, then*

$$|Pr(x \in A) - \mu(A)| \leq \mu(A)\mu(\bar{A}) \, \delta \, V_T(f).$$

Applied to n-fold iteration of the logistic map: each of the subsets Δ_i has measure 2^{-n}. This yields the desired convergence result for the parabola gadget, regarding distributions of the variable x.

Corollary 4.1.2 *Let $\{X_k\}$ be a sequence of random variables related by the logistic map (4.9). Let X_0 have a distribution that has density f with respect to the invariant measure μ, with finite total variation $V_T(f)$. Then*

$$|Pr(X_n \in A) - \mu(A)| \leq \frac{\mu(A)\mu(\bar{A})}{2^n} \, V_T(f).$$

We now prove Theorem 4.1.

Proof let χ_A be the characteristic function of A,

$$\chi_A(x) = \begin{cases} 1, & x \in A; \\ 0, & x \notin A. \end{cases} \tag{4.19}$$

Let $\lambda = \mu(A)$.

$$\begin{aligned}
Pr(x \in A) - \lambda &= \int f(x) \left(\chi_A(x) - \lambda \right) d\mu(x) \\
&= \sum_i \int_{\Delta_i} f(x) \left(\chi_A(x) - \lambda \right) d\mu(x).
\end{aligned} \tag{4.20}$$

Since $\mu(\Delta_i \cap A) = \lambda \, \mu(\Delta_i)$ for all i,

$$\int_{\Delta_i} \left(\chi_A(x) - \lambda \right) d\mu(x) = 0, \tag{4.21}$$

and so, for any numbers $\{a_i\}$,

$$Pr(x \in A) - \lambda = \sum_i \int_{\Delta_i} \left(f(x) - a_i \right) \left(\chi_A(x) - \lambda \right) d\mu(x). \tag{4.22}$$

Therefore,

$$|Pr(x \in A) - \lambda| \le \sum_i \int_{\Delta_i} \left| \left(f(x) - a_i \right) \left(\chi_A(x) - \lambda \right) \right| d\mu(x). \tag{4.23}$$

Take

$$a_i = \frac{1}{2} \left(f_i^+ + f_i^- \right). \tag{4.24}$$

Then, for almost all $x \in \Delta_i$,

$$|f(x) - a_i| \le \frac{1}{2} \left(f_i^+ - f_i^- \right), \tag{4.25}$$

and (4.23) yields,

$$|Pr(x \in A) - \lambda| \le \frac{1}{2} \sum_i \left(f_i^+ - f_i^- \right) \int_{\Delta_i} |\chi_A(x) - \lambda| \, d\mu(x). \tag{4.26}$$

Within Δ_i, the function $|\chi_A(x) - \lambda|$ is equal to $1 - \lambda$ on a set of measure $\lambda\mu(\Delta_i)$ and to λ on a set of measure $(1 - \lambda)\mu(\Delta_i)$. Therefore,

$$\int_{\Delta_i} |\chi_A(x) - \lambda| \, d\mu(x) = 2\lambda(1 - \lambda)\mu(\Delta_i). \tag{4.27}$$

We thereby obtain the result,

$$|Pr(x \in A) - \lambda| \leq \lambda(1 - \lambda)\sum_i \mu(\Delta_i)(f_i^+ - f_i^-), \tag{4.28}$$

which is what was to be proved. $\square$

We have demonstrated convergence of probability distributions for x. This means, for any measurable subset A of the unit interval, the measure of the subset of the state space that consists of all (x, z) with $x \in A$ converges towards $\mu(A)$. We will now show that we have convergence of measure, not only for such sets, but for arbitrary measurable subsets of the state space.

It suffices to show that we have convergence for rectangles $[a, b] \times [c, d]$. Now, if we partition the z-axis into 2^k bins of length 2^{-k}, then, after k iterations of the gadget's evolution, which bin z is in depends only on the initial value of x and is independent of the initial value of z. Therefore, for a rectangle of the form $[a, b] \times [p/2^k, q/2^k]$, where p and q are integers, the probability that, at any time after k iterations, the state is in that rectangle, depends only on the initial value of x. Its inverse image is the set of all points such that x is in a set D, where D has measure $\mu(D) = (b - a)(q - p)/2^k$.

We can say the same of any set that is the set of all (x, z) for $x \in A$ and $z \in L$, where A is any measurable subset of the unit interval and L is a union of intervals, of total length $\|L\|$, with endpoints that are integral multiples of 2^{-k}. The inverse image of this set under k iterations is the set of all (x, z) with $x \in D$, where D is a set of measure $\mu(D) = \rho(A \times L) = \mu(A)\|L\|$. Thus, for any n, k, $(x_{n+k}, z_{n+k}) \in A \times L$ if and only if $x_n \in D$, and

$$Pr((x_{n+k}, z_{n+k}) \in A \times L) = Pr(x_n \in D). \tag{4.29}$$

Let $\lambda = \mu(D) = \rho(A \times L)$. Then

$$|Pr((x_{n+k}, z_{n+k}) \in A \times L) - \lambda| = |Pr(x_n \in D) - \lambda| \leq \frac{\lambda(1 - \lambda)}{2^n} V_T(f), \tag{4.30}$$

where f, once again, is the density function, with respect to μ, for x_0. This gives us the following convergence result for sets of this form.

Theorem 4.2 *Let A be any measurable subset of the unit interval, and let L be a subset of the unit interval that is a union of intervals with endpoints that are integral multiples of 2^{-k}. Let $\lambda = \rho(A \times L)$. Then, for $n \geq k$,*

$$|Pr((x_n, z_n) \in A \times L) - \lambda| \leq 2^k \frac{\lambda(1 - \lambda)}{2^n} V_T(f),$$

where f is the density function for x_0 with respect to μ.

We get convergence of the distribution of z by taking A to be the unit interval.

Corollary 4.2.1 *Let L be a subset of the unit interval that is a union of intervals with endpoints that are integral multiples of 2^{-k}, and let λ be the total length of L. Then, for $n \geq k$,*

$$|Pr(z_n \in L) - \lambda| \leq 2^k \frac{\lambda(1 - \lambda)}{2^n} V_T(f),$$

where f is the density function for x_0 with respect to μ.

If we coarse-grain both u and z into 2^k bins of equal width 2^{-k}, then, for any initial distribution such that f has bounded total variation, the probability of u_n being in a given bin approaches 2^{-k} as n increases, as does the probability of z_n being in a given bin, but we have tighter bounds on the probability of u_n, by a factor 2^k.

It is easy to extend this convergence result to arbitrary intervals $L = [a, b]$, to get bounds on the rate of convergence of the distribution of z. For any interval, and any k, we can find intervals $L' = [a', b']$, $L'' = [a'', b'']$, whose endpoints are integral multiples of 2^{-k}, such that $a'' \leq a \leq a'$, $b' \leq b \leq b''$, $a' - a'' \leq 2^{-k}$ and $b'' - b' \leq 2^{-k}$. Since $L' \subseteq L \subseteq L''$, we must have

$$Pr(z_n \in L') \leq Pr(z_n \in L) \leq Pr(z_n \in L''). \qquad (4.31)$$

Let λ, λ', and λ'' be the lengths of L, L', and L'', respectively. λ' and λ'' are both within 2^{-k} of λ. We have, from (4.31),

$$Pr(z_n \in L') - \lambda' - (\lambda - \lambda') \leq Pr(z_n \in L) - \lambda$$
$$\leq Pr(z_n \in L'') - \lambda'' + (\lambda'' - \lambda). \qquad (4.32)$$

and hence

$$-\left(|Pr(z_n \in L') - \lambda'| + 2^{-k}\right) \leq Pr(z_n \in L) - \lambda$$
$$\leq |Pr(z_n \in L'') - \lambda''| + 2^{-k}. \qquad (4.33)$$

We can apply Corollary (4.2.1) to L' and L'' ; by taking k and n sufficiently large, we can get bounds as tight as we want on $|Pr(z_n \in L) - \lambda|$.

5

Epistemic Chances, or "Almost-Objective" Probabilities

5.1 Is chance incompatible with determinism?

As we saw in §2.5, on the grounds of presumed determinism of the laws of nature, a number of writers on probability, including Bernoulli, Laplace, de Morgan, and Jevons, adopted an official conception of probability that is wholly epistemic. There is a tension in these writers, however; they tend to slip into treating of it as a matter of fact, not subjective opinion, whether, for example, a given die is a fair one. And, indeed, this seems unavoidable; could it be that it really makes no sense to ask whether a given gambling device is fair or biased?

Of course, it is possible that, at bottom, the laws of physics are not deterministic. But questions such as whether we should think that the fundamental laws are deterministic get us into vexed topics such as the interpretation of quantum mechanics. These considerations seem far removed from the mundane question facing a casino operator of whether or not a given pair of dice is biased.

There is, therefore, a need for a notion that can at least act in some ways like objective chance, that does not presuppose indeterministic fundamental physics. The considerations of the previous chapter point towards a conception of this sort. The dynamics of the parabola gadget are deterministic. The dynamics are also chaotic, and small uncertainties concerning initial conditions rapidly become large uncertainties about the conditions at later times. However, from this chaos order arises, as there is a probability distribution, determined by the dynamics of the device, towards which a wide range of probability distributions over initial conditions converge. While the uncertainty about initial conditions may be epistemic, the convergence of probabilities about initial conditions to the attractor distribution is a matter of the dynamics of the device; both the fact of convergence and the character of the attractor distribution are facts about the dynamics of the device. In this

Beyond Chance and Credence. Wayne C. Myrvold, Oxford University Press (2021).
© Wayne C. Myrvold.
DOI: 10.1093/oso/9780198865094.003.0005

chapter we introduce a hybrid notion of probability, which goes beyond the familiar dichotomy of chance and credence, which is appropriate to physical set-ups of this sort.

5.2 The method of arbitrary functions

The parabola gadget provides an example of a system for which a variety of input distributions lead to virtually the same probabilities for certain claims one might make for states of affairs at other time. Don't confuse this sort of convergence with another sort, namely, the convergence towards agreement of differing credence-functions that can be brought about via Bayesian updating on a sufficiently rich body of evidence. This is completely different; there is no new information being obtained; what is at issue is the relation between probabilities about the state of the system at one time and probabilities about its state at some other time that are imposed by the dynamics of the system. As Engel (1992: 4) puts it, this is not a case of data swamping priors; rather "*the physics swamps the prior.*"

The study of this phenomenon was pioneered by von Kries (1886), Poincaré (1896), and Hopf (1934; 1936); for the history, see von Plato (1983). There are a number of mathematical results along these lines that have come to be known, following Hopf (1934; 1936), as the "method of arbitrary functions." This name is a bit misleading, as not all of the results along these lines hold for *completely* arbitrary probability density functions over initial conditions—for example, though we showed, in the last chapter, that, for the parabola gadget, any probability distribution that has a density with respect to the invariant distribution converges towards it, the interesting bounds on rates of convergence hold only for density functions of bounded total variation.

Poincaré discusses these matters in *Calcul des Probabilités*, and in two of his popular works: in chapter 11 of *Science and Hypothesis* (1902) (which is a reprint of Poincaré 1899), and in chapter 4 of *Science and Method* (1908), which was included in the second edition (1912) of *Calcul des Probabilités* as an Introduction.

Poincaré's prime example involves a roulette-like wheel, divided into a large, even number n of sectors of equal width, alternately coloured red and black, with a pointer fixed to its support. It is spun with some imperfectly known initial angular velocity, and eventually comes to rest through friction, after completing several revolutions. The question to be asked is: what is the probability that, when it comes to rest, the pointer will point to a red sector?

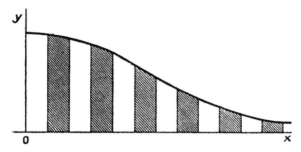

Figure 5.1. Poincaré's illustration of a density function over final angle. From Poincaré (1896: 128; 1912: 149)

It is supposed that the initial position is perfectly known; the result then depends only on the law of friction and on the probability distribution of the initial angular velocity. This induces a probability distribution over the final position. Suppose that the probability for the initial angular velocity has a density function ϕ; on reasonable assumptions about the force law, the probability distribution for the total angle the wheel rotates through before coming to rest will have a density f, which, of course, depends on ϕ. His illustration of a representative function of this sort is shown in Figure 5.1. In his treatment, Poincaré holds the input probability distribution, and hence the output probability density function f, fixed, and considers the limit as the number of sectors increases, and consequently the angle subtended by each sector decreases, without limit. As long as f is a continuous function, the probabilities for red and black converge to one-half in this limit.

This limit would correspond to a sequence of wheels with ever-increasing numbers of sectors. Of course, if one is interested in knowing how to bet on a particular wheel, set spinning by a person or mechanism with a certain limited degree of control over the initial angular velocity, this limit is the answer to the wrong question; one wants to know, for that wheel, the relation between distributions over initial and final conditions, and get bounds on how far the probability of red can deviate from one-half in terms of some feature of the distribution over initial conditions.

This is something that Poincaré does on the way to proving his limit theorem.[1] Poincaré considers the case in which f has a bounded derivative:

$$|f'(x)| < M. \tag{5.1}$$

---

[1] There's a lesson here. Mathematicians often take limits because they like nice clean results. If it is bounds on behaviour on the way to the limit that one wants, these can often be extracted from proofs of convergence.

Suppose that each sector subtends an angle ε, and that the largest angle the wheel can rotate through is A. Then, under the assumption (5.1), it follows (Poincaré 1896: 128–9; 1912: 149–50) that

$$|Pr(Red) - Pr(Black)| < MA\varepsilon. \qquad (5.2)$$

Thus, as long as the input distribution has a density that varies sufficiently slowly that the output density varies little over angles of size ε, the probabilities for red and black will be roughly equal.

This reasoning, like the reasoning employed in the previous chapter, takes probabilities as inputs, applies physical dynamics to them, and yields probabilities as outputs. To know how to interpret these results, that is, to understand how to take the output probabilities, we need to know the status of the input probabilities.

What Poincaré says about how he conceives of the input probabilities is less helpful than one would like. In *Science and Hypothesis*, he writes,[2]

> I do not know what the probability is that the pointer will be spun with a force such that this angle will lie between θ and $\theta + d\theta$. However, I can adopt a convention. I can suppose that this probability is $\phi(\theta)d\theta$. As for the function $\phi(\theta)$, I may select it in an entirely arbitrary manner, since there is nothing to guide me in my choice. However, I am naturally led to suppose that this function is continuous.
>
> (Poincaré 2018: 137, from Poincaré 1902: 234)

In chapter 4 of *Science and Method*, we find,

> What is the probability that the impulse has this or that value? About this I know nothing, but it is difficult for me not to admit that the probability is represented by a continuous analytic function.
>
> (Poincaré 1908: 78; 1912: 12)[3]

In the same breath, Poincaré treats the input probability as something about which one can be ignorant, which suggests it is something objective and

---

[2] Note that there is a shift of notation between *Calcul des Probabilités* and *Science and Hypothesis*; ϕ is now a density function over the final angle, i.e. the function we have been calling f.

[3] "Quelle est la probabilité pour que cette impulsion ait telle ou telle valeur? Je n'en sais rien, mais il m'est difficile de ne pas admettre que cette probabilité est representée par une fonction analytique continue."

independent of us, and as something that we can make arbitrary conventions about. This, it must be admitted, is confusing.

Savage (1973) suggested that we take the input probabilities to be subjective credences. One might think that this renders the output probabilities subjective. This isn't true, for reasons that we have seen in the previous chapter. If we take the input probabilities to be subjective, then the output probabilities are what result from taking these and evolving them using physical dynamics. What we start out with might be a subjective matter, but what happens to them under dynamical evolution is a matter of physics. For example: Alice and Bob might both, mistakenly, think that a given roulette wheel is fair, and hence attribute equal credence to each sector as the ball's final resting place. Unbeknownst to them, the wheel is not fair, and the result of evolving their credences about initial conditions is actually a probability distribution that favours some sectors over others. The quantities that result from evolving their credences about initial conditions via the actual dynamics might not be equal to, or even close to, their credences about states of affairs at a later time.

5.3   Epistemic chances

The probability distributions that result from taking credence-functions and evolving them according to physical dynamics don't fit neatly into the familiar dichotomy of credence and objective chance. They are hybrids. They have an epistemic aspect, in the input distributions. But what happens to the input distributions is a matter of the dynamics of the system. If there is a probability distribution towards which a wide range of input distributions converge, this is a matter of the dynamics of the system, and what that probability distribution is—that is, what values it ascribes to various propositions about the system—is determined by the dynamics of the system. We have seen this already in the previous chapter, in our discussion of the parabola gadget. The fact that probability distributions over initial values of the position of the ball tend towards μ rather than a distribution uniform in x or any other distribution is a fact about the dynamics that lead from one stage of the evolution to another.

In previous work I have borrowed a term from Schaffer (2007), and called these probabilities *epistemic chances*. The term is meant to capture the fact that there is an epistemic aspect to them, and also the fact that they are capable of playing much the same role as objective chances.

A few words about how this conception is to work.

The key feature is that we are dealing with a dynamical system whose evolution washes out differences between various probability functions over conditions at a given time. Thus, it is not only capable of reducing differences between credences of different agents, but also of producing more-or-less determinate credences from indeterminate ones. Recall our discussion back in §2.2.4. There we pointed out that it would be absurd to pretend that agents have credences defined to arbitrary precision. One way to handle this is to represent a belief set by a *credal set*, that is, a set of credence functions that the agent regards as reasonable, given what she knows. This set itself may be a bit vague, but let us restrict ourselves for the moment to the lower tier of Good's hierarchy. If you know very little about the initial state of a parabola gadget, we will represent your epistemic state by a set C of credence functions.

This set will not contain absolutely every possible function. This means that we are invoking a more stringent notion of reasonableness than mere probabilistic coherence; recall the discussion at the end of §2.2.1.

Back to the parabola gadget. If we take it that reasonable agents should not change their credences about the ball being in a given interval $[a, b]$ very much when the boundaries of the interval change by a small amount, we can implement this constraint by taking C to contain only probability distributions represented by continuous functions with derivative bounded by some quantity M. The total variation of any such function cannot be larger than M, and so the class of all such functions exponentially converge towards each other as they look further into the future. From Corollary 4.1.2 in the previous chapter, we have, for any measurable subset A,

$$|Pr(x_n \in A) - \mu(A)| \leq \frac{\mu(A)\mu(\bar{A})M}{2^n} \leq \frac{M}{2^{n+2}}. \tag{5.3}$$

We will never reach a point, no matter how large n is, at which all functions in the set C agree precisely. But the range of probabilities will narrow, and may eventually reach a point at which differences don't matter. In this way, the dynamics of a device such as the parabola gadget can generate relatively precise credences from imprecise ones.

To sum up: suppose we have

- a physical system S, with a dynamical map T_t over its state space.
- a class C_{t_0} of credence functions, to be regarded as the class of functions a reasonable agent could have, at time t_0, about the state of S at t_0.

- a threshold ε below which differences in credences are regarded as inconsequential.

Then, if there is a time t_1 such that the system's evolution between t_0 and t_1 is given by T_t and, for some proposition A about the state of the system at time t_1, every function in $\mathcal{C}_{t_0}$, when evolved via T_t from t_0 to t_1, yields a probability that is within ε of some common value p, we will say that, at time t_0, p is an *epistemic chance* of A obtaining at time t_1.

Obviously, these conditions for there being an epistemic chance of a given proposition are fairly stringent, and there will be lots of propositions for which they are not fulfilled, and hence lots of propositions that will not be assigned an epistemic chance. Nonetheless, I claim that they do hold for games of chance and in statistical mechanics and, indeed, in any situation that we can model by classical physics and in which we can rightly claim that there are well-defined objective probabilities.

5.4 On the reasonable credence-set $\mathcal{C}_t$, and an analog of the Principal Principle

In the characterization of epistemic chances is invoked a class $\mathcal{C}_t$ of credence functions, the ones to be regarded as those that a reasonable agent could have about the state of the system at time t. Objective Bayesians would regard this as a singleton, but there is neither need nor reason to be so restrictive. Differences among members of $\mathcal{C}_t$ may be due to differences in the information that agents actually have, or to differences between judgments of plausibility. The notion is meant to be friendly to approaches, already mentioned in §2.2.4, that represent states of belief by credal sets; on such an approach, one would take $\mathcal{C}_t$ to be the union of all reasonable credal sets. A minor modification of the notion would allow $\mathcal{C}_t$ itself to be an imprecise set.

If $C(\cdot)$ is any credal function in $\mathcal{C}_t$, and if E is any evidence or information possessed by some agent, then, clearly, $C(\cdot|E)$ is also in $\mathcal{C}_t$, as it cannot be unreasonable to conditionalize on information in your possession. But the idea of the reasonable credences set $\mathcal{C}_t$ is that the credences not contained in it are unreasonable, in the sense that they pretend to deeper knowledge of the system than the agent could have. For that reason it makes sense to include, for every $C(\cdot)$ in $\mathcal{C}_t$, also $C(\cdot|E)$, where E is any *accessible* evidence, meaning, evidence that could be obtained by the sorts of means available to the agent at time t. This would include, for example, any results of observations that could feasibly have been performed, or measurements that could feasibly have been made.

Our judgments about what evidence is accessible are based on our judgments about what measurements and observations could feasibly be performed. These are shaped by experience and may change with experience. A shift in judgments about the extent of the class $\mathcal{C}_t$ may occasion a shift in judgments about which propositions have epistemic chances.

What evidence is accessible depends on the time. At the present time I can measure the current value of some parameter, but can only predict its future values. If it is the case that I could have made a measurement in the past, then I could have a record of the result of that past measurement. This asymmetry of epistemic access introduces a temporal asymmetry into the class $\mathcal{C}_t$. Although we have memories and records of the past, usually the best we can do when it comes to prediction of the behaviour of physical systems is to take the current state and forward-evolve it. This, in turn, can be traced to causal asymmetry; there are processes that reliably produce records of past events; there is nothing comparable when it comes to future events. If we agree to include, for every credence function C in $\mathcal{C}_t$, also its conditionalization on any evidence E that is accessible at t, then empirically ascertainable matters pertaining to the past of t will have no epistemic chances that differ greatly from 0 and 1.

Suppose, now, that you have good reason to believe of some system, say, a roulette wheel prior to its being spun, that there is an epistemic chance about some aspects of its state at a later time, but you don't know what that epistemic chance is. You can entertain hypotheses about the value of that epistemic chance, such as the hypothesis that the roulette wheel is fair, or that it is biased in favour of the number 23. You can also form conditional credences, such as the chance of the ball landing on 23, conditional on the hypothesis that the roulette wheel is fair. In general, we can consider conditional credences of the form

$$Cr_t(A \mid ec_t(A) \in \Delta),$$

where $ec_t(A) \in \Delta$ is the proposition that the epistemic chance of A at time t is in the set Δ. What should these conditional credences be? Recall that to be an epistemic chance is to be the result of applying physical dynamics to any reasonable credence over initial conditions. Presumably, you regard your own credences (or everything in your credal set) as reasonable ones. If your credences about the future state of some system are not the results of taking your credences about its present state and evolving them according to the system's dynamics, then, on pain of incoherence, this could only be due to uncertainty about what the result of applying such an operation would be. Your conditional credence, conditional on the supposition that the probability ascribed to a proposition A that results from applying such an operation to your credences about the present state of the system lies in an interval Δ, should lie in Δ. Thus, your conditional credences, at time t, should satisfy

$$Cr_t(A \mid ec_t(A) \in \Delta) \in \Delta. \tag{5.4}$$

This is the analog, for credences about epistemic chances, of the Principal Principle, which applies to credences about wholly objective chances. Call it *ECPP*, for *Epistemic Chance Principal Principle*.

Since the class $\mathcal{C}_t$ is taken to be closed under conditionalization on evidence accessible at t, then it follows from (5.4) that, for any E that is accessible at t,

$$Cr_t(A \mid E \,\&\, ec_t(A) \in \Delta) \in \Delta. \tag{5.5}$$

This bears some resemblance to the original Principal Principle, which is stipulated to be robust under conditionalization on what Lewis dubbed *admissible evidence*, defined as evidence that affects credence in A only via its impact on credence about the chance of A. There is a crucial difference between (5.5) and the Principal Principle as Lewis conceived of it. Lewis takes as admissible every detail of the past, "no matter how hard it might be to discover" (Lewis 1980: 172). This has the consequence that, if the laws of physics are deterministic, there are no chances other than zero and one. Our (5.5) has no such consequence, as E is restricted to accessible evidence.

5.5 Learning about epistemic chances

As we've already stressed, epistemic chances are not credences, but the result of applying physical dynamics to credences. For this reason, an agent might have good reason to believe that a proposition *has* an epistemic chance, but might be uncertain about its value, either because of uncertainty about the dynamics of the system in question, or because the dynamics are known but she has not done the requisite calculation (which, in cases more complicated than our parabola gadget, might be intractable). She might, however, have various credences about the value of the epistemic chance.

Hypotheses about the value of an epistemic chance can be put to the test, in much the way that hypotheses concerning chances are. Suppose we have a means of generating a sequence of trials whose outcomes we believe to have equal, independent epistemic chances. We can update our credences about the values of these chances by conditionalizing on the outcomes of these trials.

For example, in Table 5.1, we present the result of simulated experiments involving parabola gadgets. For the first experiment, 1,000 initial conditions were chosen at random, with distribution uniform in x, within the interval $[0.3, 0.4]$. These initial conditions were evolved through 100 steps of the gadget's operation. The results were coarse-grained by partitioning the diagonal into 20 bins of equal length, and counting the number of outcomes in each bin. These results are tabulated in Table 5.1, and graphed as a histogram in Figure 5.2, along with expectation values (blue lines) calculated on the basis of the invariant measure μ.

Table 5.1 Results of simulated parabola gadget experiment

Bin	Measure of bin	Results in 1,000 trials	Results in 10,000 trials
[0, 0.05]	0.1436	146	1,402
(0.05, 0.10]	0.0613	62	629
(0.10, 0.15]	0.0483	50	480
(0.15, 0.20]	0.0420	46	463
(0.20, 0.25]	0.0382	38	389
(0.25, 0.30]	0.0357	34	360
(0.30, 0.35]	0.0340	32	309
(0.35, 0.40]	0.0329	38	354
(0.40, 0.45]	0.0322	28	335
(0.45, 0.50]	0.0319	26	280
(0.50, 0.55]	0.0319	32	353
(0.55, 0.60]	0.0322	33	289
(0.60, 0.65]	0.0329	24	284
(0.65, 0.70]	0.0340	29	357
(0.70, 0.75]	0.0357	35	369
(0.75, 0.80]	0.0382	37	395
(0.80, 0.85]	0.0420	47	413
(0.85, 0.90]	0.0483	49	495
(0.90, 0.95]	0.0613	77	636
(0.95, 1]	0.1436	137	1,408

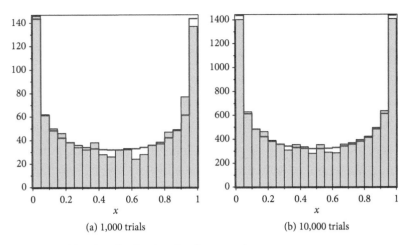

(a) 1,000 trials   (b) 10,000 trials

Figure 5.2. Results of a simulated gadget experiment

Suppose that you were convinced that the dynamics of the gadget are the right sort to produce epistemic chances for propositions of the sort "After 100 iterations, the ball is in the interval [0, 0.05]," but were uncertain about the values of those epistemic chances. Should you take results of an experiment such as the one simulated (that is, an experiment performed with actual, physical parabola gadgets) as evidence that calculations of probabilities according to the measure μ yield those epistemic chances, either exactly or at least a good approximation to them?

Let ec be the (unknown, or imperfectly known) epistemic chance function, and suppose that, for some interval A of the diagonal, you have prior credences (or a set of them) in propositions such as $ec(A) \in \Delta$, for all sub-intervals Δ of the unit interval. We assume that your conditional credences satisfy the ECPP introduced in the previous section. With this constraint on conditional credences in place, learning about epistemic chances proceeds exactly along the same lines as our account of learning about chances in §2.4. Suppose, for example, that A is the proposition that the ball ends up, after 100 iterations, within 10 cm of the right-hand edge of the gadget (Alice's option from the previous chapter). This has probability, on the measure μ, of about 0.2048. In our first experiment, 214 out of 1,000 balls ended up in A, and, in the second experiment, 2,044 out of 10,000. Conditionalizing on these results, and applying the ECPP, amounts to multiplying a prior credence density over the epistemic chance of A by functions proportional to the likelihood functions depicted in Figure 5.3. Provided that your prior

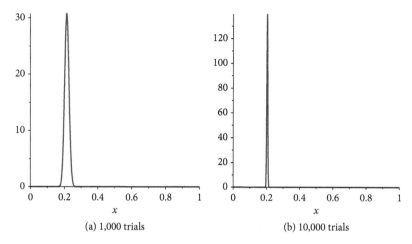

Figure 5.3. Normalized likelihood functions for a simulated gadget experiment

credences are not too dogmatic, your posteriors will end up concentrated near the observed relative frequency.

5.6 Soup from a stone: can the epistemic aspect be eliminated?

The so-called method of arbitrary functions takes probabilities as input, applies physical dynamics to them, and yields probabilities as output. If no probabilities enter in, no probabilities come out. However, the dynamics play such a significant role, and we ask so little of the input probabilities, that there is a temptation to try to modify the method so that it can yield probabilities from non-probabilistic input. Perhaps, the hope goes, in this way we could eliminate the epistemic aspect from this conception, in which case we would have, not merely something that fulfills much of the function of objective chances, but *genuinely objective chances*.

I am sympathetic to attempts to make something along these lines work. It would be terrific if the method of arbitrary functions could be made to generate probabilities from deterministic dynamics alone, or from deterministic dynamics plus something objective. Alas, I fear that this is too much like getting soup from a stone in the fable. The stone is not enough; there must be some other ingredient before we have anything that affords nourishment, and the method cannot generate probabilities *de novo*.

Michael Strevens (2003; 2011; 2013), Marshall Abrams (2012), and Jacob Rosenthal (2012) have all advanced proposals that invoke the sorts of dynamical considerations we have been considering, but which aim at using them to formulate a wholly objective conception of probability. Though, as I have said, I am sympathetic to these attempts, I don't think any of them achieve the aim.

The problem that each of these proposals has to face is the *problem of input distributions*. Rosenthal simply takes it for granted that there is some standard measure that can be thought of as a *natural* one, in the sense that distributions of initial conditions that are in accord with this measure, that is, probable on its lights, are the default assumption, needing no explanation, whereas distributions of initial conditions that aren't in accord with it would require explanation. Not much guidance is offered as to which distribution this is. It is said to be Lebesgue measure, but, on a given state space that has the topological and differential structure of $\mathbb{R}^n$, there are many distinct Lebesgue measures, corresponding to different coordinatizations of the space. There is a further danger associated with reasoning of this sort. If we expect initial conditions of physical systems to be chosen according to measures uniform in canonical phase-space variables, then we should expect everything to be in a macrostate corresponding to thermodynamic equilibrium, an expectation refuted by virtually everything that we observe, and, indeed, by the mere fact *that* we observe.

Though their proposals differ in details, Strevens and Abrams each invoke facts about actual frequencies of initial conditions of multiple copies of relevantly similar systems. There are two major problems with any such proposal. First, it requires there to be an abundance of initial conditions, whereas, *prima facie*, for systems like our parabola gadget, there is a matter of fact about whether option *A* or option *B* is a better bet, whether or not the Universe contains copies of the parabola gadget with varying initial conditions. Second, a proposal of this sort runs the danger of making the connection between chances and frequencies analytic (recall our discussion in §3.2.3). If facts about frequencies of initial conditions enter into the definition of the output probabilities, then it is *guaranteed* that the relative frequencies of various outcomes approximate their respective probabilities. But, if the concept is to be in accord with the Weak Law of Large Numbers, this close approximation should be merely *probable*. Consider again the example of our simulated experiment with the parabola gadget, with the outcome frequencies depicted in Figure 5.2. These outcome frequencies strongly support the hypothesis that the correct probabilities are those given by μ over rival hypotheses, such as the one that the correct probabilities are uniform in x, because the results are overwhelmingly more probable on the hypothesis that the μ-probabilities are the right ones than they are on the hypothesis that they are uniform. But neither hypothesis accords probability zero to *any* distribution of outputs. We may regard initial conditions that lead, say, to all outputs ending up in one bin, as extraordinarily, perhaps *absurdly* unlikely, but they should not be regarded as impossible. And that means that we need a notion of probability over initial conditions. Proposals that attempt to do away with input probability distributions, or else to define them in terms of actual frequencies, seem ill-equipped for this task.

Dynamical considerations of a similar flavour are also invoked in Hoefer's *Humean Best Systems* account of chances. It is an adaptation, to the case of chances, of the Humean Best Systems Account of laws, though Hoefer is not committed to an account of that sort for the dynamical laws of physics. The idea, insofar as I understand it, is this. Consider a complete account of basic, non-lawlike facts about the world. Such a description would be enormous, and complex. Consider, now, another description, brief enough to be graspable by a human mind, offered as a systematization of the enormous and complex complete description. Alternate systematizations might be offered. These can be ranked, it is said, in terms of simplicity, strength, and fit. The Best System is the one that scores highest in this ranking.

Hoefer's proposal is to take the laws of physics as given, and to apply the account to chances. The intuitive idea is that, though a description that specifies the outcome of every coin toss throughout history is stronger than one that merely provides probabilities, a statement of probabilities would be considerably simpler.

As in all such discussions, no precise characterization of the criteria of strength, simplicity, and fit is provided. There is, it seems, not a single account here, but a family of different accounts, one for each way of measuring these criteria. It seems to be presumed that all criteria that are in accord with our intuitions about such things will yield essentially the same system, but, in the absence of actual proposed candidates, it is hard to judge this claim.

Fit, presumably, is to be judged by some sort of accuracy score, such as the Brier score mentioned in §3.1.4, or some other proper scoring rule. Simplicity, for Hoefer, is largely a matter of user-friendliness, which is "a combination of two factors: *utility* for epistemically and ability-limited agents such as ourselves, and *confirmability*" (Hoefer 2016: 82–3).

Let us attempt to apply this to the parabola-gadget wager offered to Alice and Bob, from §4.2. A prescription to bet on A if the actual condition of the gadget is such as to put the ball in A at the time of readout, and to bet on B if the actual condition is such as to put the ball in B, would fail the user-friendliness criterion badly, as it would have no utility whatsoever for agents with the epistemic limitations of Alice and Bob.

How would a prescription to assign credences $\mu(A)$ and $\mu(B)$ to the two eventualities fare? These are the values of the epistemic chances, and, if (as Hoefer does) the method of arbitrary functions is to be invoked, they are the values yielded by that method. These values are

$$
\begin{aligned}
\mu(A) &= 1 - (2/\pi)\arcsin(\sqrt{0.9}) \approx 0.205; \\
\mu(B) &= (2/\pi)\left(\arcsin(\sqrt{0.6}) - \arcsin(\sqrt{0.4})\right) \approx 0.128.
\end{aligned}
\tag{5.6}
$$

I don't know whether an injunction to assign these credence-values to the propositions A and B is to count as user-friendly or not. It depends, I suppose, on the calculational resources available to the agents. This is not a calculation that I, personally, can do in my head, but it is an easy one if one has access to an appropriate calculating device.

The essential point is that epistemic considerations have not been eliminated from the conception. If utility for epistemically-limited agents involves considerations about what the agents could know about the physical set-up, then, it seems, we are deeply embroiled with considerations of what sorts of credences might be reasonable for epistemically limited agents in a given situation.

5.7 Conclusion

Epistemic chances, as we have defined them, do not fit into the familiar dichotomy of objective chance and subjective credence. They are hybrids. Both epistemic considerations, represented by the class of reasonable credences $\mathcal{C}$, and physical considerations, represented by the dynamical map, appear in their characterization. When an epistemic chance exists, its value is largely a matter of dynamics. This is illustrated by the case of the parabola gadget. As long as the class $\mathcal{C}$ of reasonable credences includes some credence-functions that have a density with respect to Lebesgue measure on the state space, there could be no distribution that represents epistemic chances arbitrarily far into the future other than the invariant distribution ρ.

The epistemic aspect is ineliminable; without any limitations on the range of acceptable input distributions, we get no convergence of probabilities. However, for certain kinds of systems—and these, I claim, include, by construction, devices used in games of chance, and also the sorts of systems to which we successfully apply statistical mechanics—the limitations are not terribly stringent, and the input distributions that participate in the convergence plausibly include all those that could be in the credal set of a reasonable agent.

In spite of their epistemic aspect, they fill the role that objective chances are thought to play in our lives. For a given gambling device, there may be a matter of fact about the approximate value of epistemic chances of outcomes. Moreover, these facts can be known; we can subject hypotheses regarding them to empirical test, and end up with well-informed credences in their values.

The claim that they are equipped to play the role of probabilities required in statistical mechanics will take some arguing. This I will do in Chapter 8, after a brief introduction to thermodynamics and statistical mechanics, in the next two chapters.

6

Thermodynamics: The Science of Heat and Work

6.1 About the term "thermodynamics"

In what the author describes as the "first systematic treatise on that science
which has ever appeared," Rankine (1859: 299) defines the science of thermodynamics as follows:

> It is a matter of ordinary observation, that heat, by expanding bodies, is a
> source of mechanical energy, and conversely, that mechanical energy, being
> expended either in compressing bodies, or in friction, is a source of heat. . . .
> The reduction of the laws according to which such phenomena take place,
> to a physical theory, or connected system of principles, constitutes what is
> called the SCIENCE OF THERMODYNAMICS.

Don't be misled by the term "thermodynamics," which suggests the dynamics
of heat, or, in other words, the science of the motion of heat from one place
to another. The name *thermodynamics*, a term composed out of the Greek
words for *heat* and *power*, and often hyphenated as "thermo-dynamics" in
the first decades of its existence, refers to the study of the ways in which heat
can be used to generate mechanical action and, conversely, of how heat may
be generated via mechanical means.[1]

---

[1] The word's first appearance is in pt VI of Kelvin's "On the Dynamical Theory of Heat"
(Thomson 1857, read before the Royal Society of Edinburgh on May 1, 1854). There he reca-
pitulates what in 1853 he had called the "Fundamental Principles in the Theory of the Motive
Power of Heat," now re-labelled "Fundamental Principles of General Thermo-dynamics." The
context makes clear that the new term is intended to denote the study of the relations between
mechanical action and heat.

> Mechanical action may be derived from heat, and heat may be generated from
> mechanical action, by means of forces either acting between contiguous bodies,
> or due to electric excitation; . . . Hence Thermo-dynamics falls naturally into two

Beyond Chance and Credence. Wayne C. Myrvold, Oxford University Press (2021).
© Wayne C. Myrvold.
DOI: 10.1093/oso/9780198865094.003.0006

Thermodynamics begins with the observation that heat can be generated via mechanical action such as rubbing two objects together, and that one can continually generate heat on this way, without limit, as long as you can keep the mechanical action going. It thus involves the rejection of the notion, prevalent up until the mid-nineteenth century, that heat is a conserved fluid (sometimes called *caloric*) that is neither created nor destroyed and merely flows from one object to another.

We do work by moving something against a resisting force; the quantity of work done is proportional to the resisting force and to the distance moved. Thus, raising a ten-pound weight one foot requires the same amount of work as raising a one-pound weight ten feet, and, if the two weights are coupled by a lever-arm or a pulley, the ten-pound weight can do work on the one-pound weight by decreasing its height by an amount one-tenth of the distance the one-pound weight is raised.

There is a mechanical equivalent of heat: the same amount of work always generates the same amount of heat, and, when heat is consumed to do work, the same amount of heat is always consumed to perform a given amount of work. Thus we can measure both work and heat in the same units, units of energy. Though heat is not a conserved quantity, energy is; when heat is created by friction, there is an expenditure of work equal to the heat created. This is the First Law of Thermodynamics, which Kelvin states as,[2]

Prop. I. (Joule).—When equal quantities of mechanical effect are produced by any means whatsoever by purely thermal sources, or lost in purely thermal effects, equal quantities of heat are put out of existence or are generated.   (Thomson 1853: 264; 1882: 178)

Divisions, of which the subjects are, respectively, *the relation of heat to the forces acting between contiguous bodies*, and *the relation of heat to electrical agency*.
(Thomson 1857: 123; 1882: 232)

Maxwell, in 1878, defines thermodynamics as "the investigation of the dynamical and thermal properties of bodies, deduced entirely from what are called the First and Second Laws of Thermodynamics, without any hypotheses as to the molecular constitution of bodies" (Maxwell 1878; Niven 1890: 664–5). It is worth noting that the term "thermodynamics" had not been used in connection with the earlier investigations, by Fourier, Kelvin, and others, of the laws of heat transport.

[2] In 1853, in pt I of "On the Dynamical Theory of Heat," this is referred to as "Prop. 1." It becomes "Law I" in 1857, in his recapitulation of the "Fundamental Principles of Thermo-Dynamics," as the subject was now called, in pt VI (erroneously labelled "Part V" in the original journal publication, even though there had already been a pt V in the previous volume). In Rankine (1859) we finally find a "First Law of Thermodynamics" and a "Second Law," though Rankine's Second is not what we call the Second Law of Thermodynamics.

Energy is always conserved (thus, admonishments to conserve energy are easily obeyed), but it can be *dissipated*. Dissipation of energy involves a lost opportunity to do work. Two paradigm examples of dissipation are

    I. Flow of heat from one body to another at a lower temperature.
    II. Free expansion of a gas into a chamber at a lower pressure.

For the latter, imagine a gas initially confined by a partition to one part of a vessel that is otherwise empty. The partition is removed, and the gas expands to fill the volume now available to it, and settles into an equilibrium in which it fills the vessel at uniform pressure.

In both cases, there is a lost opportunity to do work. If heat flows from a hot body to a cooler one, we could have used that heat to a drive a heat engine. A heat engine operates by extracting heat from a higher-temperature reservoir, using it to expand a gas, which can drive a piston or turbine, and then recompressing the gas, discarding the heat into a lower-temperature reservoir. If we simply let the same amount of heat flow from one reservoir to another, without doing any work, then, since some of that heat could have been converted to work, there is a lost opportunity. The same goes for free expansion of a gas; the expansion could have been harnessed to move a piston, which could raise a weight.

In any process there is some dissipation. In order for a heat engine to extract heat from a reservoir, there must be heat flow between the reservoir and the working substance of the engine, and this requires a temperature difference. But, if we are patient, and are willing to do it slowly, this can be done in small stages, with less dissipation. It is common, and has been since the early days of thermodynamics, to imagine an ideal limiting process in which no dissipation occurs. A common device in the literature, with its roots in Carnot (1824), is the reversible engine, one that can be worked forwards as an engine to produce mechanical work while extracting heat from a high-temperature source, discarding part of it into a low-temperature sink, or backwards as a refrigerator, using mechanical work from an external source to remove heat from a cooler reservoir and transport it to a higher-temperature reservoir, with the same efficiency in either direction. The workings of the reversible engine are instances of a more general category of *reversible processes*.

Despite its ubiquity in thermodynamics texts, the concept of a reversible process is, as John Norton (2016) has recently argued, not as clear as one would like. In order for a process to occur at all, there must be *some* dissipation of energy; otherwise, nothing happens. However, though there can be no dissipationless process, it is assumed that we can achieve any state transition with dissipation as small as we like. Norton proposes that we take the properties ascribed to the nonexistent and impossible reversible processes to be limiting properties of *sets* of realizable processes. For instance: it is an established part of thermodynamics that any two reversible heat engines operating between a heat source and a heat sink of given temperatures have the same efficiency, which is therefore a function only of the temperatures of source and sink, and that any irreversible engine has a lower efficiency. The physical content of this can be expressed without reference to a fictitious reversible engine: given any two temperatures, there is an efficiency that is an upper bound that the efficiency of no engine operating between heat source and sink at those temperatures can either equal or exceed, though efficiencies of physically possible engines can come arbitrarily close to it.

6.2 The Laws of Thermodynamics

Kelvin identified two fundamental laws of thermodynamics. The First we have already seen; the Second we will come to shortly. However, another law has been recognized, regarded as more fundamental than the first two, which has accordingly come to be known as the Zeroth Law. There is a law yet more fundamental than all of these, which has been called the Minus First Law (see Brown and Uffink 2001 for extended discussion). There is also a Third Law of Thermodynamics (which, though important, will play no role in our discussions, and will be ignored).

6.2.1 The Zeroth Law of Thermodynamics

The concept of thermodynamic equilibrium can be used to introduce the concept of temperature. Two bodies that are in thermal contact (this means that heat flow between them is possible), which are in thermal equilibrium with each other, will be said to have the *same temperature*. We want this relation of equitemperature to be an equivalence relation. It is reflexive and symmetric by construction. That it is transitive is a substantive assumption[3] (though one that has often been taken for granted). Suppose we have bodies

---

[3] Noted and emphasized by, among others, Maxwell; see Maxwell (1871: 32–3).

that can be moved around and brought into thermal contact with each other. When this happens, the contact might induce a change of state (brought on by heat transfer from one to the other), or it might not. The Zeroth Law says that, if the states of two bodies A, B are such that they are in equilibrium with each other when brought into thermal contact, and the states of B and C have the same relation, then the states of A and C are such that they are in equilibrium with each other when brought into thermal contact.

6.2.2 The First Law of Thermodynamics

The First Law of Thermodynamics is aptly named, even though the Zeroth Law is more basic. It is the law that makes thermodynamics possible, as it says that work and heat can be measured on the same basis. There is a mechanical equivalent of heat; in any process in which work is converted entirely into heat, the same amount of work always produces the same amount of heat, no matter the nature of the process. This means that, for any body, we can define a total internal energy U, which is a function of the state of the system, such that the change in internal energy U is a sum of the heat flow (positive for heat transferred into the system, negative for heat out) and the work done on or by the system (a positive quantity for work done on the system, corresponding to a net increase in energy, negative for work done by the system on the external world). In differential form (corresponding to vanishingly small increments of energy), this is written,

$$dU = \dbar Q + \dbar W. \tag{6.1}$$

The increments of heat, $\dbar Q$, and of work, $\dbar W$, are written with bars to emphasize that, unlike the change in U, these do not represent changes in some property of the state of the system. There is no such thing as the total heat content of the system, or total work content; there is only the total energy, which has two modes of change: via work done on or by the system, or via heat transfer.[4]

---

[4] Note that the terminology "heat transfer" is a relic of the old way of thinking about things, on which heat was thought to be a substance that could flow from one body to another.

6.2.3 The Second Law of Thermodynamics

Kelvin's Second Fundamental Principle of Thermodynamics is

> Prop II. (Carnot and Clausius).—If an engine be such that, when it is worked backwards, the physical and mechanical agencies in every part of its motion are all reversed, it produces as much mechanical effect as can be produced by any thermo-dynamic engine, with the same temperatures of source and refrigerator, from a given quantity of heat.
> (Thomson 1853: 264; 1882: 178)

He does not, however, take this as a primitive axiom; its foundation, he says, is founded on the axiom that has come to be known as the *Kelvin statement of the Second Law of Thermodynamics.*

> *It is impossible, by means of inanimate material agency, to derive mechanical effect from any portion of matter by cooling it below the temperature of the coldest of the surrounding objects.*   (Thomson 1853: 265; 1882: 179)

Clausius had earlier argued that no two reversible engines operating between the same source and sink could have different efficiency, because, if there were such engines, by using one to drive the other as a refrigerator,

> ...it would be possible, without any expenditure of force or any other change, to transfer as much heat as we please from a *cold* to a *hot* body, and this is not in accord with the other relations of heat, since it always shows a tendency to equalize temperature differences and therefore to pass from *hotter* to *colder* bodies.   (Clausius 1899: 90, from Clausius 1850: 503)

At the time, this was not singled out as a special principle. A few years later it became an alternate form of the Second Law of Thermodynamics.

> Heat can never pass from a colder to a warmer body without some other change, connected therewith, occurring at the same time.
> (Clausius 1856: 86, from 1854: 488; 1864: 134)

> A passage of heat from a colder to a hotter body cannot take place without compensation.   (Clausius 1879: 78, from Clausius 1876: 82)

This is known as the *Clausius statement of the Second Law*. It is prefaced by some discussion of the sorts of processes to be counted as compensatory.

The striking feature of these statements, and the one that has garnered most attention, is their temporal asymmetry. Both declare to be impossible some processes whose temporal inverse is evidently *not* impossible. The temporal inverse of the process that Clausius' statement deems impossible is spontaneous heat transfer from a hotter to a cooler body, which is, of course, the normal course of things. The temporal inverse of the process deemed impossible in the Kelvin statement would be a process in which one begins with a body that is cooler than any surrounding body and, by exerting mechanical effect on it, warms it enough so that it is no longer cooler than its surroundings. This is also readily done.

Both statements of the Second Law rely on a distinction between heat and work. The processes they declare impossible are not impossible if one replaces *heat* by *mechanical effect*, and vice versa. Though there is no passage of heat from a cooler body to a hotter without compensation, a cooler body can do work on a hotter without compensation; a thermally insulated, cool gas at high pressure can expand and raise a warm weight. The Kelvin statement also relies on a distinction between mechanical effect and work. It is not impossible to cool a body to a temperature below that of its surroundings (refrigerators exist). What the Kelvin statement says is that we will have to expend more work to do so than is derived from the body in the process; that is, there cannot be a positive net mechanical effect of the process. But, of course, it is *not* true that, in the process of cooling a body, more heat has to be put into the body than is extracted from it!

Thermodynamics is aptly named the science of heat and work. The key insight behind the First Law is the interconvertibility of the two, and, without the distinction, neither the Clausius statement nor the Kelvin statement of the Second Law can be formulated.

6.3 The Equilibration Principle

Another principle, more basic than the Zeroth Law, has been identified as a law of thermodynamics. This is what Brown and Uffink (2001) call the *Equilibrium Principle* (though perhaps *Equilibration Principle* would be better):

An isolated system in an arbitrary initial state within a finite fixed volume will spontaneously attain a unique state of equilibrium.

(Brown and Uffink 2001: 528)

To signal that this is more fundamental than the traditional Zeroth, First, and Second Laws, Brown and Uffink label this the *minus first law*. They point out that a principle of this sort had been recognized as a law of thermodynamics earlier, by Uhlenbeck and Ford (1963: 5).[5]

The equilibration principle has often been conflated with the Second Law. The two are distinct, however. It is a consequence of the Second Law that, *if* an isolated system makes a transition from one thermodynamic state to another, the entropy of the final state will not be lower than that of the initial state. This does not entail that its behaviour will be that dictated by the equilibration principle. As far as the Second Law is concerned, there might be some set of distinct thermodynamic states of the same entropy that the system cycles through. Or else it might fail to equilibrate when isolated, remaining in some quasistable non-equilibrium state until some external disturbance triggers a slide towards equilibrium.

Compared to the other laws of thermodynamics, a law to the effect that, left to themselves, systems tend to relax to equilibrium, is an outlier. Though long recognized as an important principle, it was a late entry to the list of laws of thermodynamics: it was not referred to as a law of thermodynamics until the 1960s, more than a century after Kelvin initiated talk of laws, or fundamental principles, of thermodynamics. There is also an important conceptual distinction between the equilibration principle and the other laws of thermodynamics. It is unique among the so-called laws of thermodynamics in that its formulation does not require a distinction between energy transfer as heat and energy transfer as work.

Though, of course, this is ultimately a matter of choice of terminology, and nothing more, it seems to me that it is helpful to highlight the differences between the equilibration principle and the more traditional laws of thermodynamics by restricting the scope of "thermodynamics" to something like what its founders intended, and to regard the equilibration principle, not as belonging to thermodynamics proper, but as a *presupposition* thereof.

---

[5] Uhlenbeck and Ford called it the *Zeroth Law* to emphasize its priority over the First and Second Laws. We will continue to follow standard terminology in taking transitivity of thermal equilibrium to be the Zeroth Law.

A payoff in conceptual clarity of this choice is that it will perhaps mitigate somewhat the tendency to conflate the equilibration principle with the Second Law.

6.4 Enter entropy

In a Carnot cycle (the process undergone by the reversible engine imagined by Carnot), a quantity of heat Q_h is taken in by the engine from a reservoir at a temperature T_h, some of which is converted into work, and the remainder, Q_c, discarded into a reservoir at a lower temperature T_c. By the First Law, the work done is

$$W = Q_h - Q_c. \tag{6.2}$$

If we define the *efficiency* η of the engine as the fraction of Q_h that is converted to work, that is, as W/Q_h, we have

$$\eta = \frac{W}{Q_h} = \frac{Q_h - Q_c}{Q_h} = 1 - \frac{Q_c}{Q_h}. \tag{6.3}$$

It is a consequence of the Second Law that the efficiency of any reversible engine operating between reservoirs at temperatures T_h and T_c is the same. Following a suggestion of Joule, Kelvin proposed to use this to define an absolute temperature scale (Thomson 1848). Define the ratio of the temperatures T_c and T_h to be

$$\frac{T_c}{T_h} = \frac{Q_c}{Q_h} = 1 - \eta. \tag{6.4}$$

Then we have, by definition,

$$\frac{Q_h}{T_h} = \frac{Q_c}{T_c}. \tag{6.5}$$

In a Carnot cycle in the forward direction, heat Q_h flows into the engine, and Q_c flows out; when the engine is operated in reverse, the directions of heat flow are reversed. For any cycle, let Q_1 be the heat transferred between engine and reservoir at temperature T_h, taken to be a positive quantity for heat flow into the engine, and negative for heat flow out of it, and similarly for Q_2. This means that, for the Carnot cycle in the positive direction, $Q_1 = Q_h$ and

BEYOND CHANCE AND CREDENCE 131

$Q_2 = -Q_c$, and, for the reverse cycle, $Q_1 = -Q_h$ and $Q_2 = Q_c$. For either cycle,

$$\frac{Q_1}{T_h} + \frac{Q_2}{T_c} = 0. \tag{6.6}$$

Clausius (1854) argued that this must hold for any reversible cycle, on the grounds that, if it didn't, cycles could be chained to produce a cycle whose net effect was delivery of heat from a colder reservoir to a hotter, with no net expenditure of work. We can consider more complicated cycles, with more than two reservoirs; the same sort of reasoning entails that, for any reversible cycle,

$$\sum_i \frac{Q_i}{T_i} = 0. \tag{6.7}$$

For an irreversible cycle, in which a system is restored to the same thermodynamic state in which it started, we have,

$$\sum_i \frac{Q_i}{T_i} < 0. \tag{6.8}$$

If the heat is transferred, not at constant temperature, but at continuously varying temperature, we can replace the sum by an integral, and say that, for a reversible cycle,

$$\int \frac{dQ}{T} = 0. \tag{6.9}$$

This is what Clausius calls "an analytic expression of the second *Hauptsatz* of the mechanical theory of heat, valid for all reversible cyclic processes" (Clausius 1854; 1864: 147).

For an irreversible cyclical process,[6]

$$\int \frac{dQ}{T} < 0. \tag{6.10}$$

---

[6] In 1854 Clausius has the opposite sign, but that's because he's adopted the opposite sign convention regarding heat flow; Clausius was then taking heat flowing out of the engine as positive. In Clausius (1865) (the paper in which he introduces the term *entropy*), and subsequent works, he adopts the convention still in use.

In 1854 Clausius called the quantity Q_i/T_i the *equivalence-value* of the heat transferred at temperature T_i. This later (Clausius, 1865) became *entropy*.[7]

If two thermodynamic states a, b can be connected by a reversible process, then the quantity

$$\int_a^b \frac{dQ}{T} \tag{6.11}$$

will have the same value for *any* such process. We can thus define a function S of the thermodynamic state (defined up to an arbitrary constant), the *entropy*, such that the entropy difference of two states

$$S_b - S_a = \int_a^b \frac{dQ}{T}, \tag{6.12}$$

where the integration is to be taken along any reversible process connecting the two states.

The change of entropy between the initial and final states of any process, as long as these are states of thermodynamic equilibrium, is well-defined, even if the process is not a reversible one, as long as the states can be connected by a reversible process. What you do, to calculate the entropy difference between

---

[7] Here's what Clausius says about the rationale for the new term.

If we look for a designation for S, one might, in analogy to saying that the quantity U is the body's *heat and work content*, say of the quantity S that it is the *transformational content* of the body. But since I think it better to take the names of such important scientific quantities from ancient languages, so that they can be used unchanged in all modern languages, I propose that S be called according to Greek words, ἡ τροπή, *transformation*, the *entropy* of the body. The word *entropy* was deliberately formed to be as similar as possible to the word *energy*, because the two magnitudes which are to be named by these words are so closely related to one another in their physical meanings, that a certain similarity in the designation seems to me to be appropriate.

"Sucht man für S einen bezeichnenden Namen, so könnte man, ähnlich wie von der Grösse U gesagt ist, sie sei der *Wärme- und Werkhinhalt* des Körpers, von der Grösse sagen, sie sei der *Verwandlungsinhalt* des Körpers. Da ich es aber für besser halte, die Namen derartiger für die Wissenschaft wichtiger Grössen aus den alten Sprachen zu entnehmen, damit sie unverändert in allen neuen Sprachen angewandt werden können, so schlage ich vor, die Grösse S nach dem griechischen Worte ἡ τροπή, die Verwandlung, die *Entropie* des Körpers zu nennen. Das Wort *Entropie* habe ich absichtlich dem Wort *Energie* möglichst ähnlich gebildet, denn die beiden Grössen, welche durch diese Worte benannt werden sollen, sind ihren physikalischen Bedeutungen nach einander so nahe verwandt, dass eine gewisse Gleichartigkeit in der Benennung mir zweckmässig zu sein scheint."

(Clausius 1865: 390; 1867a: 34)

states a and b, is cook up *some* reversible process connecting the two states, and apply (6.12). The Second Law guarantees that it will not matter which reversible process you choose.

For example, consider one of our paradigm examples of an irreversible process, free expansion of a gas. Suppose that a gas is enclosed in an insulating container, and is initially confined by a partition to a volume V_i. The partition is removed, and the gas expands to fill the entire volume V_f. A reversible process that leads to the same final state could involve, for example, a piston that is slowly moved out, with the force on the piston nearly balancing the pressure exerted on it by the gas. The gas does work on the piston, and hence would lose energy without compensating input of energy. In the original, irreversible process that we're trying to mimic, the gas exchanges no energy with its environment, and so its initial and final energies are the same. We achieve this final state reversibly by coupling the gas to a heat bath of the same temperature as the gas.[8] As the gas is slowly expanded, the energy expended as work is replaced by heat absorbed from the heat bath. The entropy difference between initial and final states is the amount of heat absorbed by the gas during this process, divided by the temperature. It is positive, since heat *enters* the gas. Therefore, free expansion of a gas is a process in which its entropy increases.

This will be a general feature of irreversible processes taking place in systems that are isolated from the external world and exchange no heat with it. Here's the argument. Since, for any process (6.10) holds, it must be the case that the integral in (6.12) takes its maximum possible value, among all processes that connect a and b, on a reversible process. That is, for any process in which heat may flow in and out of a system, as long as we can ascribe a temperature to the system at all stages of the process, we must have,

$$\int_a^b \frac{dQ}{T} \leq S_b - S_a. \tag{6.13}$$

This holds for all processes whatsoever, reversible or irreversible, taking place in isolated or non-isolated systems. If the system is isolated, however, it

---

[8] For a sufficiently rarified gas, the temperature is constant as long as the energy is constant (this is Joule's law), in which case a heat bath of fixed temperature suffices; otherwise, adjust the temperature of the heat bath to match, at every stage of the process, the instantaneous temperature of the gas.

exchanges no heat with its environment, and the integral on the left is zero. Therefore, we have, for any process taking place in an isolated system,

$$\Delta S = S_f - S_i \geq 0. \tag{6.14}$$

6.5 Thermodynamics reconceived

The interconvertibility of heat and work lends credibility to an idea that had been proposed earlier, namely, that heat is a form of motion of microscopic parts of matter, and most of those who accepted the interconvertibility also eventually accepted the kinetic theory of heat. The kinetic theory of heat, also known as the *dynamical theory of heat* or the *mechanical theory of heat*, has an interesting consequence: the laws of mechanics are applicable to thermal phenomena. This, in turn, requires a reconceptualization of the Second Law of Thermodynamics. If an isolated physical system is subject to the laws of mechanics, this has far-reaching consequences. For one thing, if the force laws are invariant under velocity reversal (as they will be if intermolecular forces satisfy Newton's third law and depend only on inter-molecular distances), then, for any process that is consistent with the laws of mechanics, its time-reversal will be, also, and thus, a monotonic relaxation to equilibrium cannot be a consequence of dynamics alone (this is the *reversibility argument*). Second, for an isolated system confined to a bounded region of space, there can be no function of the state that monotonically increases; this was urged, as an objection to the kinetic theory, by Zermelo and Poincaré (see Brown et al. 2009 for discussion). These facts require a reconceptualization of the Second Law of Thermodynamics. The Second Law, as originally conceived, cannot be strictly true, and it must be replaced by something of a probabilistic character.

6.5.1 "Dear Strutt,—": the reversibility argument

The reversibility argument first appears in a pencilled annotation on the letter from Maxwell to P. G. Tait (Dec. 11, 1867) that contains the earliest exposition of the "very observant and neat-fingered being" that would come to be known as *Maxwell's demon*. Tait had asked Maxwell for advice concerning his *Sketch of Thermodynamics* (published 1868), and one of Maxwell's suggestions was "To pick a hole—say in the 2nd law of Θ^{cs}...." Maxwell then provides an exposition of the demon thought-experiment, concluding, in short if heat is the motion of finite portions of matter and if we can apply tools to such portions of matter so as to deal with them separately then we can take advantage of the different motion of different portions to restore a uniformly hot system to unequal temperatures or to motions of large masses.

Only we can't, not being clever enough.
(Knott 1911: 214; Garber et al. 1995: 177; Harman 1995: 332)

On this letter someone[9] wrote, "Very good. Another way is to reverse the motion of every particle of the Universe and to preside over the unstable motion thus produced" (Knott 1911: 214; Garber et al. 1995: 178; Harman 1995: 332; Harman 2002: 186). The argument is spelled out in a letter, dated April 7, 1868, from Maxwell to the editor of the *Saturday Review* (Garber et al. 1995: 187–8; Harman 1995: 360–61). We find a more expansive exposition of the argument in a letter dated Dec. 6, 1870, from Maxwell to John William Strutt, Baron Rayleigh.

Dear Strutt,—

If this world is a purely dynamical system, and if you accurately reverse the motion of every particle of it at the same instant, then all things will happen backwards to the beginning of things, the raindrops will collect themselves from the ground and fly up to the clouds, etc, etc, and men will see their friends passing from the grave to the cradle till we ourselves become the reverse of born, whatever that is. We shall then speak of the impossibility of knowing about the past except by analogies taken from the future, but I do not think it requires such a feat to upset the 2nd law of thermodynamics.
(Garber et al. 1995: 205; Harman 1995: 582)

Maxwell follows this with an exposition of the now-familiar demon, and then draws the

Moral. The 2nd law of thermodynamics has the same degree of truth as the statement that if you throw a tumblerful of water into the sea, you cannot get the same tumblerful of water out again.

---

[9] The pencilled annotation is attributed to Kelvin by C. G. Knott, editor of *Life and Scientific Work of Peter Guthrie Tait*, and this attribution has been widely accepted. However, a more recent examination has called this into doubt; Garber, Brush, and Everitt write, "The present editors consider the writing closer to Tait's than to Thomson's" (Garber et al. 1995: 178).

What the Second Law, as originally conceived, declared to be impossible, Maxwell now regarded as but highly improbable.

We find similar considerations in Gibbs several years later,

> when such gases have been mixed, there is no more impossibility of the separation of the two kinds of molecules in virtue of their ordinary motions in the gaseous mass without any external influence, than there is of the separation of a homogeneous gas into the same two parts into which it has once been divided, after these have once been mixed. In other words, the impossibility of an uncompensated decrease of entropy seems to be reduced to improbability.   (Gibbs 1875: 229; 1906a: 167)

A consequence of the reversibility argument is that it would be futile to try to derive the Second Law from the laws of mechanics alone. In a letter to Tait of 1873, three years before Loschmidt urged reversibility considerations upon Boltzmann, Maxwell ridiculed attempts to found the Second Law on principles of mechanics.

> But it is rare sport to see those learned Germans [Clausius and Boltzmann] contending for the priority of discovering that the 2nd law of $\Theta\Delta^{cs}$ is the Hamiltonsche Princip, ... The Hamiltonsche Princip, the while, soars along in a region unvexed by statistical considerations, while the German Icari flap their waxy wings in nephelococcygia amid those cloudy forms which the ignorance and finitude of human science have invested with the incommunicable attributes of the Queen of Heaven ...
> (Knott 1911: 115–16; Garber et al. 1995: 225; Harman 1995: 947).[10]

Boltzmann's probabilistic turn occurred around 1877. In 1872 Boltzmann had proven a theorem, which has come to be called the H-theorem, which gave the appearance of deriving, from mechanical considerations alone, a conclusion that the distribution of velocities in a gas always monotonically approaches a Maxwell–Boltzmann distribution (the generalization of the Maxwell distribution to a gas subjected to an external force, such as gravity), independently of the initial state of the gas. This is, of course, impossible, on reversibility grounds. It was via a paper of Loschmidt (1876) that Boltzmann's attention was drawn to reversibility considerations, though criticism

---

[10] Note that Maxwell's abbreviation of "thermodynamics" is a nod to the Greek roots of the word.

of the H-theorem was not Loschmidt's purpose in the paper (see Darrigol 2018: 181–8 for discussion). In his response (1877a), Boltzmann acknowledged that his conclusion that the gas equilibrates must be regarded as only probable, and other sorts of behaviour must be regarded as improbable but not impossible.

Thus, in the decade from 1867 to 1877, the major figures involved in the development of statistical mechanics all concluded, on the basis of the reversibility argument, that the Second Law of Thermodynamics, as originally conceived, could not be strictly true, and that it must be replaced by a probabilistic version, in which what is deemed impossible in the original version becomes improbable. This probabilistic version of the Second Law is what is meant by physicists today, when they speak of the Second Law of Thermodynamics.

6.5.2 Equilibrium reconceived

> In a system in thermodynamic equilibrium, the parameters (e.g. the energy content of a subsystem) vary in time—they fluctuate.
>
> (Szilard 1972: 70, translation of 1925: 753)

In thermodynamics, equilibrium is thought of as a state in which nothing is happening. On the molecular level, however, the stillest waters are seething with activity, as molecules bounce and jostle each other. The apparent calm is due to the limitations of our powers of observation.

Considerations of goings-on at the molecular level entail that, even though the properties of a system in equilibrium are stable when considered macroscopically, at the molecular scale quantities such as the density of matter within a given region will be fluctuating. And, it turns out, these fluctuations are observable, the most vivid and best-known example being the perpetual dance of pollen grains and other small particles suspended in a liquid, known as Brownian motion.

Though one could not expect to predict the precise trajectory of a Brownian particle, we can treat the surrounding medium statistical-mechanically and calculate a probability distribution over observable quantities such as the displacement of the Brownian particle over a given period of time, and it is found that observed frequency distributions accord well with the calculated probabilities.

Consider, now, a ball that is projected horizontally into a viscous medium, at some substantial initial velocity. Friction with the medium will slow the ball's motion until it has lost its horizontal velocity. The kinetic energy of the ball is not destroyed, but, rather, is transferred to the medium as heat, warming it somewhat. This counts as an instance of equilibration.

Now let the ball be small enough that it has observable Brownian motion. The state into which it settles is not one in which it has lost all its horizontal velocity, but rather, one in which it bounces haphazardly one way or another, with no preferred direction with respect to the surrounding medium. This process of settling down into a situation that is not quiescent with respect to all measurable variables but is, rather, a condition characterized by a stable pattern of fluctuations, is every bit as much a process of equilibration as is a process in which the end condition is one in which all observable variables are constant. From the point of view of statistical mechanics, this is how equilibrium is to be characterized: as a stable probability distribution that unstable ones tend towards. Note that there are two shifts that are made. One is that equilibrium is to be characterized, not in terms of a stable state of the system, but in terms of a stable pattern of fluctuations, naturally conceived in terms of a stable probability distribution over these fluctuations. The second is that equilibrium is a condition that a system relaxes *towards*, not a condition reached in a finite time, though it may, after a finite time, reach a condition imperceptibly different from the condition it approaches.

6.5.3 The Second Law as a statistical regularity

As we saw in Chapter 1, Maxwell saw the Second Law as a statistical regularity of the same sort as the statistical regularities that social scientists were finding, in abundance, in human populations.

> The data of the statistical method as applied to molecular science are the sums of large numbers of molecular quantities. In studying the relations between quantities of this kind, we meet with a new kind of regularity, the regularity of averages, which we can depend upon quite sufficiently for all practical purposes, but which can make no claim to that character of absolute precision which belongs to the laws of abstract dynamics.
>
> (Maxwell 1873b: 440; Niven 1890: 374)

The truth of the second law is . . . a statistical, not a mathematical, truth, for it depends on the fact that the bodies we deal with consist of millions of molecules, and that we never can get hold of single molecule.

> (Maxwell 1878: 279; Niven 1890: 670)

If the Second Law is a statistical regularity, this means that it is not exceptionless, though exceptions will be negligible when we are dealing with bulk samples of matter containing large numbers of molecules. On the scale of molecular processes things are otherwise.

If we restrict our attention to any one molecule of the system, we shall find its motion changing at every encounter in a most irregular manner.

If we go on to consider a finite number of molecules, even if the system to which they belong contains an infinite number, the average properties of this group, though subject to smaller variations than those of a single molecule, are still every now and then deviating very considerably from the theoretical mean of the whole system, because the molecules which form the group do not submit their procedure as individuals to the laws which prescribe the behaviour of the average or mean molecule.

Hence the second law of thermodynamics is continually being violated, and that to a considerable extent, in any sufficiently small group of molecules belonging to a real body. As the number of molecules in the group is increased, the deviations from the mean of the whole become smaller and less frequent; and when the number is increased till the group includes a sensible portion of the body, the probability of a measurable variation from the mean occurring in a finite number of years becomes so small that it may be regarded as practically an impossibility.

This calculation belongs of course to molecular theory and not to pure thermodynamics, but it shows that we have reason for believing the truth of the second law to be of the nature of a strong probability, which, though it falls short of certainty by less than any assignable quantity, is not an absolute certainty.   (Maxwell 1878: 280; Niven 1890: 670–71)

In the letter to Rayleigh, quoted above, the lesson drawn by Maxwell from the demon is that

In this way the temperature of B may be raised and that of A lowered without any expenditure of work, but only by the intelligent action of a mere guiding agent (like a pointsman on a railway with perfectly acting switches who should send the express along one line and the goods along another). I do not see why even intelligence might not be dispensed with and the thing made self-acting.

(Garber et al. 1995: 205; Harman 1995: 583)

Maxwell introduces the demon to the public in his *Theory of Heat*, in a section entitled "Limitation of the Second Law of Thermodynamics."

One of the best established facts in thermodynamics is that it is impossible in a system enclosed in an envelope which permits neither change of volume nor passage of heat, and in which both the temperature and the pressure are everywhere the same, to produce any inequality of temperature or pressure without the expenditure of work. This is the second law of thermodynamics, and it is undoubtedly true as long as we can deal with bodies only in mass, and have no power of perceiving the separate molecules of which they are made up. But if we conceive of a being whose faculties are so sharpened that he can follow every molecule in its course, such a being, whose attributes are still as essentially as finite as our own, would be able to do what is at present impossible to us. For we have seen that the molecules in a vessel full of air at uniform temperature are moving with velocities by no means uniform, though the mean velocity of any great number of them, arbitrarily selected, is almost exactly uniform. Now let us suppose that such a vessel is divided into two portions, A and B, by a division in which there is a small hole, and that a being, who can see the individual molecules, opens and closes this hole, so as to allow only the swifter molecules to pass from A to B, and only the slower ones to pass from B to A. He will thus, without expenditure of work, raise the temperature of B and lower that of A, in contradiction to the second law of thermodynamics.

This is only one of the instances in which conclusions which we have drawn from our experience of bodies consisting of an immense number of molecules may be found not to be applicable to the more delicate observations and experiments which we may suppose made by one who can perceive and handle the individual molecules which we deal with only in large masses.   (Maxwell 1871: 308–9)

Note that there is in this no hint that there might be some principle of physics that precludes the manipulations of the demon, or constrains it to dissipate sufficient energy that the net change of entropy it produces is positive. Moreover, Maxwell leaves it open that the requisite manipulations might become technologically possible in the future—the demon does what is *at present* impossible for us. What Maxwell is proposing, as a successor to the Second Law, is strictly weaker than the probabilistic version. For Maxwell, even the probabilistic version is limited in its scope—it holds only in circumstances in which there is no manipulation of molecules individually or in small numbers.

Thus, for Maxwell, there is a two-fold limitation on the Second Law, as originally conceived. First, it is merely a statistical regularity, holding with high probability as long as matter composed of a great many molecules is being considered, but continually violated at the molecular level. Second, just as statistical regularities in human populations could be altered by, for example, selective interventions on individual people, the Second Law would not hold even as a statistical regularity in the presence of an agency, intelligent or otherwise, that could selectively manipulate individual molecules.

6.5.4 The Maxwellian view: distinction between heat and work as means-relative

The kinetic theory of heat not only requires a reconceptualization of the Second Law; it requires us to rethink the fundamental distinction at the heart of thermodynamics, the distinction between energy transfer as heat and energy transfer as work.

To do work on a body is to exert a force on parts of it and thereby change the state of motion of those parts. On the kinetic theory, the same can be said of the process of heating a body. The difference is that, in the former case, there is a change of motion visible on the macroscopic scale; large parts of the body move in concert. When we heat a body, on the other hand, the motion of its individual molecules is altered in a haphazard way. This suggests that the distinction has to do with our ability to keep track of motions, and, indeed, Maxwell said so, in his review of the second edition of Tait's *Sketch of Thermodynamics* (1877).

> The second law relates to that kind of communication of energy which we call the transfer of heat as distinguished from another kind of communication of energy which we call work. According to the molecular theory the only difference between these two kinds of communication of energy is that the motions and displacements which are concerned in the communication of heat are those of molecules, and are so numerous, so small individually,

and so irregular in their distribution, that they quite escape all our methods of observation; whereas when the motions and displacements are those of visible bodies consisting of great numbers of molecules moving all together, the communication of energy is called work.

(Maxwell 1878: 279; Niven 1890: 669)

Dissipation of energy involves a reduction in its usefulness, and this suggests that the distinction has to do with *our ability* to use the energy. Both Maxwell and Gibbs took as an example the interdiffusion of two initially separated samples of gas at the same temperature and pressure. Does this represent an entropy increase, or not? The standard answer is: if the two gases are the same—two samples of ordinary air, for example—then no change in the thermodynamic state has taken place, and there has been no entropy increase. If, however, they are different, then, yes, entropy has increased, by an amount known as the *entropy of mixing*. The relevant criterion of distinctness is whether or not the two gases could be separated again by a reversible process, with some expenditure of work. If so, that process could have been run in reverse, to produce work, and the interdiffusion was a lost opportunity to obtain useful work. If we were to discover a hitherto unknown difference between gases from two sources, of the sort that could be so exploited, then, says Maxwell, "the process of interdiffusion which we had formerly supposed not to be an instance of dissipation of energy would now be recognized as such an instance." He continues,

It follows from this that the idea of dissipation of energy depends on the extent of our knowledge. Available energy is energy which we can direct into any desired channel. Dissipated energy is energy we cannot lay hold of and direct at pleasure, such as the energy of the confused agitation of molecules which we call heat. Now, confusion, like the correlative term order, is not a property of material things in themselves, but only in relation to the mind which perceives them. A memorandum-book does not, provided it is neatly written, appear confused to an illiterate person, or to the owner who understands thoroughly, but to any other person able to read it appears to be inextricably confused. Similarly the notion of dissipated energy could not occur to a being who could not turn any of the energies of nature to his own account, or to one who could trace the motion of every molecule and seize it at the right moment. It is only to a being in the intermediate stage, who can lay hold of some forms of energy

while others elude his grasp, that energy appears to be passing inevitably
from the available to the dissipated state.

(Maxwell 1877: 220–221; Niven 1890: 646)

If we lacked the ability to turn any energy to our advantage, then all processes
would be to us like the interdiffusion of identical gases: no opportunity
to do work lost, and hence no dissipation of energy, and no increase of
entropy. If, on the other hand, we had supernatural abilities of manipulation,
no energy would ever be lost to us, and, again, we would not regard any
process as dissipatory. Consider, once again, one of our paradigm examples
of dissipation, the free expansion of a gas. If a gas is confined by a partition
to one side of a chamber, with a vacuum on the other side, and if you know
this, and if you have the ability to replace the partition with a moveable
piston, you can use your knowledge of the state of the gas to do work. If
the gas is permitted to freely expand to fill the entire chamber, this is a lost
opportunity to do work. You still (provided that the gas is isolated from
external influences) know quite a bit about the state of the gas, namely, that it
is in some state that can evolve from a state in which the gas is confined to one
side of the chamber. But this is of no use to you. If, however, you knew the
position of every molecule, and were able to position a mini-piston in front
of each of them—or, to adapt Kelvin's picture (1874), a vast army of demons
with cricket bats that reverse the velocities of all the molecules—, you could
derive substantial mechanical effect from the gas. A being with such abilities
would not regard the free expansion of the gas as a lost opportunity to do
work, and would not regard the free expansion as an increase in entropy.

Something similar was expressed by Clausius (1877), in response to the
hole picked in the Second Law by Tait, in his *Lectures on Some Recent
Advances in Physical Science*, via invocation of Maxwell's demon (Tait 1876:
118–20; see also Tait 1877: 37). Responding to Tait's (unfair) charge that the
fact that the possibility of a demon that could, without expenditure of work,
cool a body below the temperature of its surroundings "is absolutely fatal to
Clausius' reasoning," Clausius wrote,[11]

---

[11] "Dieses kann ich in keiner Weise zugeben. Wenn die Wärme als eine Molecularbewegung
betrachtet wird, so ist dabei zu bedenken, dass die Molecüle so kleine Körpertheilchen sind,
dass es für uns unmöglich ist, sie einzeln wahrzunehmen. Wir können daher nicht auf einzelne
Molecüle für sich allein wirken, oder die Wirkungen einzelner Molecüle für sich allein erhalten,
sondern haben es bei jeder Wirkung, welche wir auf einen Körper ausüben oder von ihm
erhalten, gleichzeitig mit einer ungeheuer grossen Menge von Molecülen zu thun, welche
sich nach allen möglichen Richtungen und mit allein überhaupt bei den Molecülen vorkom-
menden Geschwindigkeiten bewegen, und sich an der Wirkung in der Weise gleichmässig
betheiligen, dass nur zufällige Verschiedenheiten vorkommen, die den allgemeinen Gesetzen

This I can in no way concede. If heat is regarded as a molecular motion, it should be remembered that the molecules are parts of bodies that are so small that it is impossible for us to perceive them individually. We can therefore not act on single molecules by themselves, or obtain effect from individual molecules by themselves, but rather, in every action that we exert on a body or receive from it, we have simultaneously to do with an immensely large collection of molecules, which move in all possible directions and with all the speeds occurring among the molecules, and participate in the action uniformly, in such a way that there occur only random differences, which are subject to the general laws of probability. This circumstance forms precisely the characteristic property of that motion which we call heat, and on it depends the laws that distinguish the behavior of heat from that of other motions.

If now demons intervene, and disturb this characteristic property by distinguishing between the molecules, and molecules of certain speeds are permitted passage through a partition, molecules of other speeds refused passage, then one may no longer regard what happens under these conditions as an action of heat and expect it to agree with the laws valid for the action of heat.   (Clausius 1877: 132)

A consequence of this way of thinking about the work-heat distinction is that the Second Law itself exhibits the same means-relativity as the distinction between heat and work. There are as many versions of the Second Law as there are ways to draw the distinction between heat and work, and a process that counts as a violation of a version of the Second Law predicated on one way of the drawing the distinction will not be a violation on another. What a Maxwell demon (if such a thing were possible) could effect is something that would count, on anything like the usual way of drawing the work/heat distinction, as passage of heat from a cooler to a warmer body without expenditure of work. But, as Clausius observes, the demon would at the same time prompt a revision in our conception of the controllable aspects of the system on which it operates, with the upshot that

der Wahrscheinlichkeit unterworfen sind. Dieser Umstand bildet gerade die charakteristische Eigenthümlichkeit derjenigen Bewegung, welche wir Wärme nennen, und auf ihm beruhen die Gesetze, welche das Verhalten der Wärme von dem anderer Bewegungen unterscheiden.

Wenn nun Dämonen eingreifen, und diese charakteristische Eigenthümlichkeit zestören, indem sie unter den Molecülen einen Unterschied machen, und Molecülen von gewissen Geschwindigkeiten den Durchgang durch eine Scheidewand gestatten, Molecülen von anderen Geschwindigkeiten dagegen den Durchgang verwehren, so darf man das, was unter diesen Umständen geschieht, nicht mehr als eine Wirkung der Wärme ansehen und erwarten, dass es mit den für die Wirkungen der Wärme geltenden Gesetzen übereinstimmt."

we would no longer regard the energy transfer effected by the demon as a transfer of heat.

6.6 Thermodynamics as a resource theory

The Maxwellian view of thermodynamics has undergone a resurgence in recent years. This resurgence is inspired by quantum information theory. Quantum information theory is the study of the ways in which agents can exploit various resources, such as quantum entanglement, to perform tasks that could not be reliably performed without such resources. This field can be thought of as a subfield of a broader field of inquiry: *resource theories.*

A theory of this sort invokes a background physical theory, say, quantum mechanics. It then posits agents with a specified set of operations they can perform, and a specified task to attempt. The theory studies how certain physical resources (entanglement, or thermodynamic nonequilibrium), or information about the system, can be exploited to achieve better performance at the specified task than could be achieved without these resources. A theory of this sort involves physics, of course, because it is physics that tells us what the effects of specified operations will be. But not *only* physics; considerations not contained in the physics proper, having to do with specification of which operations are to be permitted to the agents, are brought to bear. See del Rio et al. (2015) for a general framework for resource theories, Wallace (2016b), Bartolotta et al. (2016), and Gour et al. (2019) for some recent work. See Myrvold (2020d) for further discussion, and defense, of this way of thinking about thermodynamics. For reviews, see Goold et al. (2016) and Lostaglio (2019).

7
Statistical Mechanics: The Basics

7.1 Introduction

Statistical mechanics deals with macroscopic systems that, at the microphysical level, possess a large number of degrees of freedom and evolve in such a complicated manner that it would be unthinkable to attempt to track the microphysical evolution of the system. It takes a cue from the insight, mentioned above, that, on the kinetic theory of gases, events such as the spontaneous separation of gases that have been permitted to mix should be regarded, not as impossible, but merely (for macroscopic gases) extraordinarily improbable. This means that we will have to treat such systems in probabilistic terms. Among the goals of statistical mechanics is to account for the laws of thermodynamics insofar as they are valid, and to delimit their domains of validity, but the subject is not restricted in its scope to thermal behaviour. The statistical mechanical literature today consists of an ever-expanding tool-box of techniques that are applied with substantial success. There is, however, a lack of consensus on the rationale for these techniques.

In the literature on the philosophy of statistical mechanics, it is common to distinguish two main approaches, referred to as "Boltzmannian" and "Gibbsian." This terminology is misleading, as both have their roots in the work of Boltzmann. They are often presented as incompatible rivals, rather than complementary approaches. This is not how those who developed the subject saw it; Gibbs (1902), and Einstein (1902; 1903), who independently developed much of the apparatus developed by Gibbs, saw themselves as building upon and extending the work of Boltzmann. They were right about this, and, in my opinion, the divergence between "Gibbsian" and "Boltzmannian" approaches has been exaggerated (more on this in §7.5).

Both approaches deal with the evolution of probability functions on the state spaces of physical systems, so I will begin with saying something about that. I will, for the most part, deal with *classical* statistical mechanics in this chapter. However, much of what is said applies *mutatis mutandis* to quantum

Beyond Chance and Credence. Wayne C. Myrvold, Oxford University Press (2021).
© Wayne C. Myrvold.
DOI: 10.1093/oso/9780198865094.003.0007

mechanics, and it is possible, in many cases, to write down formulas that can be interpreted as valid formulas of either classical or quantum mechanics. Quantum theory will be the focus of Chapter 9.

7.2 Evolution of probability functions under classical dynamics

As we have remarked, given the law of evolution of a dynamical system, there will be a corresponding evolution of a probability measure (or any other measure) over its state space. If we have a probability measure P_0 over states of the system at some time, which we will call $t = 0$, then the probability that at some time t its state will be in a given region A of its state space, provided that it evolves undisturbed in the interim, is equal to the probability that at $t = 0$ it is in some state that evolves into a state in A.

Let T_t be the mapping of the state space that represents the system's evolution from time 0 to time t, which may or may not be invertible. For any subset A of the state space, let $T_t^{-1}(A)$ be the inverse image of A, that, is, the set of states that evolve into states in A (note that this makes sense even if the dynamics are not invertible, that is, even if some distinct initial states get mapped to the same final state). Then, given a measure μ_0, we define

$$\mu_t(A) = \mu_0(T_t^{-1}(A)). \tag{7.1}$$

If, for all measurable sets A, $\mu_t(A)$ is the same for all t, we will say that the measure is a *stationary*, or *invariant* measure. If, for all measurable A, and all t, $\mu_0(T_t(A)) = \mu_0(A)$, we will say that the measure μ_0 is *conserved*. If the mapping is invertible, these are equivalent; a measure is conserved if and only if it is invariant. In the general case, being conserved is a strictly stronger condition. Any conserved measure is invariant, but, if invertibility is not assumed, the converse may not hold.[1]

---

[1] Think of the parabola gadget's mapping of the diagonal to itself. This is not invertible, as each point on the diagonal, with the sole exception of the point on the extreme right, can be reached from two distinct points. The measure that, in Chapter 4, we called μ, is an invariant measure. But it is not conserved. The left half of the diagonal, which has μ-measure $1/2$, gets mapped onto the whole of the diagonal, which has μ-measure 1. If we consider, not just the mapping from the diagonal to itself, but rather, the evolution on the whole state space of the gadget, comprising both the position x on the diagonal and the supplementary variable z, the dynamical map on this state space is invertible, and the measure we called ρ is both invariant and conserved.

Consider a system with state space with an invariant measure μ, confined to a region of state space that has finite total measure. For such a system, it is easy to show that two things of interest for statistical mechanics hold. The first is that the integral, with respect to μ, of any measurable function over the entire accessible region of state space is constant. This means that, if the function increases along some trajectories, this increase must be compensated by a decrease along other trajectories. The second is the *Poincaré recurrence theorem*: In any region A of state space with non-zero measure, all points in A, with the exception of at most a set of measure zero, eventually return to A (though not necessarily all at the same time).[2] Proof of the recurrence theorem can be found in the Appendix to this chapter.

In classical mechanics, the state of a system consisting of a number n of point particles may be specified by specifying the positions and velocities of all of the particles with respect to some reference frame. One can also specify positions and momenta, and (provided that the masses are specified), freely go back and forth between these two coordinatizations of the space. There, are, of course, other possibilities; a popular choice is to specify the position and velocity of the center of mass of the system, and the positions and velocities of the particles with respect to the center of mass. For a rigid body, the state can be specified by specifying the position and velocity of some representative point, and appropriate angle variables, and their rates of change, to indicate its orientation and how it changes.

There is a formulation of classical mechanics that is particularly well suited to studying the evolution of probability functions, and that is the *Hamiltonian formulation*. One first writes the state in terms of *canonical variables*,

---

[2] Poincaré's proof is found in an article on the three-body problem, published in 1890; see Poincaré (1966) for English translation of the relevant parts. The theorem is also found in ch. 12 of Gibbs (1902: 139–41). Gibbs makes no mention of Poincaré or anyone else as a source for the theorem. This raises the question of whether he arrived at it independently, or simply neglected to provide a citation. I find independent discovery more plausible. Gibbs was already thinking about the implications of Liouville's theorem for statistical mechanics as early as 1884. In September of that year he gave a talk at the meeting of the American Association for the Advancement of Science entitled "On the Fundamental Formula of Statistical Mechanics, with Applications to Astronomy and Thermodynamics." The text of the talk was not published, but the abstract was (Gibbs, 1885). In it we find that the fundamental formula of the title is the Liouville equation, which is our eq. (7.3). It is a direct consequence of this that Liouville measure is invariant. Once we have that, the proof of the recurrence theorem, as you can see in the Appendix to this chapter, is only a few lines, and involves no difficult concepts. It would not be surprising if Gibbs arrived at the recurrence theorem independently of Poincaré. On the other hand, it *would* be surprising if he spent several years thinking about the applications of the Liouville theorem to statistical mechanics *without* realizing that it entails the recurrence theorem.

which consist of a set of coordinates and their associated *conjugate momenta*. If the coordinates are spatial positions, the momenta are the usual momenta, that is, velocities multiplied by masses. If the coordinates are angular coordinates, their conjugate momenta are angular momenta. A strength of the Hamiltonian formulation is its friendliness to coordinate changes. The form of the equations of motion is unchanged under a class of coordinate transformations known as *canonical transformations*, and in many cases the key to solving a problem in Hamiltonian mechanics is to find a set of coordinates in which the problem becomes a trivial one. The state space of a system, parameterized by canonical variables, is called its *phase space*.

The *Hamiltonian function* is a function that, given a point in the phase space of a system, returns the total energy of the system in a state represented by that point. Typically, this will be a sum of a kinetic energy term and a term representing the potential energy associated with the system's configuration.

So, suppose we have a system whose state is specified by canonical coordinates $\{q_i\}$ and their conjugate momenta $\{p_i\}$, and suppose we have a Hamiltonian that is a function of these and, perhaps, also of time. Then *Hamilton's equations of motion* are,

$$\frac{dq_i}{dt} = \frac{\partial H}{\partial p_i}; \quad \frac{dp_i}{dt} = -\frac{\partial H}{\partial q_i}. \qquad (7.2)$$

Given canonical coordinates, we can define a measure on phase space, on which the measure of a $2n$-dimensional "rectangle," with sides of length $\Delta q_i, \ldots, \Delta q_n, \Delta p_i, \ldots, \Delta p_n$ is the product of the lengths of all these sides. This measure extends, via countable additivity, to the Borel subsets of the phase space, which is the smallest σ-algebra containing all of these rectangles.

This measure, which is called the *Liouville measure* on phase space, has two cool features.

1) It doesn't matter which set of canonical variables we use to define it; they all yield the same measure.
2) It is conserved under Hamiltonian evolution (7.2), for *any Hamiltonian* whatsoever.

It should be emphasized: Liouville measure is invariant even if the Hamiltonian is time-dependent. This is important because we will sometimes want to consider systems that are subjected to external influences, such as, for example, electric or magnetic fields, but are otherwise isolated from their environment, and, in particular, are thermally insulated from other systems. In such a case, the Hamiltonian of the system may change with time if the values of these fields are changing, but the Liouville measure is still conserved, even if the total energy of the system is not conserved.

Because of cool feature 2, it is convenient, whenever possible, to represent a probability measure on phase space via a density function with respect to this invariant measure. If ρ is a probability density with respect to the Liouville measure, its changes as the system evolves are like the density of an incompressible fluid flowing about. This flow is captured by the *Liouville equation*,

$$\frac{\partial \rho}{\partial t} + \sum_{i=1}^{N} \left(\frac{\partial \rho}{\partial q_i} \frac{\partial H}{\partial p_i} - \frac{\partial \rho}{\partial p_i} \frac{\partial H}{\partial q_i} \right) = 0. \qquad (7.3)$$

If the density ρ is a function only of some quantity A that is conserved under the evolution of the system—that is, if ρ takes on the same value on all phase points that share the same value of A—then the probability distribution represented by that density is an invariant one. In particular, if the Hamiltonian is a time-independent function, then energy is conserved, and any distribution that is a function only of the energy is a stationary distribution.

The theorems we have mentioned, about systems with an invariant measure, were of some historical importance in the development of Boltzmann's thinking about statistical mechanics. Consider an isolated system that is confined to a region of phase space with finite Liouville measure. It could, for example, be a system confined to a box of finite spatial volume, with an upper bound on its energy. For such a system, any function that never decreases along any trajectory must have a constant, unchanging value along all but a set of measure zero of trajectories. This means that it would be futile to try to find a function of the physical state of a system that, for an isolated system, can increase but can never decrease, as entropy was thought to do, on the original conception of the laws of thermodynamics. This point was raised by Zermelo (1896), against Boltzmann's claim, in his H-theorem, to have derived monotonic entropy increase from the principles of mechanics (see Brown et al. 2009 for discussion). It also means that the Poincaré recurrence theorem applies. Incredible as it may seem, the theorem says that a box of gas, if left to evolve, free of any outside influences, would, having first evolved from a nonequilibrium macrostate to its equilibrium macrostate, eventually return to a state indistinguishable from its initial state, for all initial states that equilibrate, with the exception of at most a set of measure zero initial states.

The take-away message of this section: there is a measure, the Liouville measure, such that, as long as a system is isolated from other systems, is an invariant measure. Furthermore, if a probability distribution is represented by a density with respect to this measure that is a function only of some conserved quantity, that distribution is also invariant under evolution.

7.3 Gibbsian statistical mechanics

7.3.1 Equilibrium distributions: canonical, microcanonical, and grand canonical

Gibbs (1902) identified certain probability distributions on phase space as ones of particular interest. Following Boltzmann (1871a),[3] and in the spirit of the then prevalent frequentism about probabilities, Gibbs thought of these probability distributions in terms of a hypothetical collection of systems, which he called an *ensemble*, but nothing in the approach is wedded to frequentism about probabilities; all that is needed is that we can make sense of probability talk. In what follows, we will (for the time being) be neutral about the interpretation of these probability distributions, and, where Gibbs talked about canonical, microcanonical ensembles, etc., we will talk about *distributions*.

Consider a system that is in thermal contact with a heat bath at a fixed temperature. As a result of its interactions with the heat bath, its state will be changing in an unpredictable way, but there is reason for optimism that there is a stable distribution that initial probability distributions will tend to relax into. For such a situation, it is to be expected that the total energy of the combined system, consisting of the system of interest plus the heat bath, will be dominated by the energy of the system plus the energy of the heat bath, with interactions between the two being so small compared to these as to make a contribution negligible in size though not in effect.[4] For such a system, Gibbs suggested what he dubbed the *canonical distribution*, whose

---

[3] See also Boltzmann (1898: §32; 1964: 297–300).

[4] Note to philosophers: when someone refers to some quantity as negligible compared to some other quantity, resist the temptation to replace this with talk of an idealized situation in which that quantity is zero. If there were no interactions between the system of interest and the heat bath, there would be no tendency for the system to come into thermal equilibrium with the heat bath. What is being required is that the interaction term be small compared to the internal energies of the system of interest and the heat bath.

density function with respect to Liouville measure is, within the accessible region of the system's phase space,

$$\rho(x) = Z^{-1} e^{-\beta H(x)}, \tag{7.4}$$

where Z is a normalization constant, and β is $1/kT$, where T is the absolute temperature, and k is a constant, known as *Boltzmann's constant*. The qualification "within the accessible region of the phase space" is there to take into account the fact that the system might, for instance, be a gas confined to a box, which constrains the positions of all its component molecules to lie within that box (some restriction of this sort is required, in order that the distribution be normalizable).

The canonical distribution has some features appropriate to the situation imagined. For one thing, it is a stationary distribution, and, for another, a combined system composed of parts that are independently canonically distributed at the same temperature will, provided that the total Hamiltonian is approximately the sum of the component Hamiltonians, be itself canonically distributed. Furthermore, if two such systems are coupled via an interaction Hamiltonian, permitting energy to be transferred between them, the expectation value of energy flow will be from the higher temperature system to the lower.

It can be argued that a canonical distribution is uniquely well suited to represent a thermal system.[5] The idea is that a thermal system is one such that no work can be done on, or by it. In the statistical context, this requires modification. Due to random fluctuations, we may be able to obtain work from the system. But an attempt to do so might cost us. We impose as a condition on a distribution representing a thermal system, that the *expectation value* of the work obtained from the system be zero, no matter how we manipulate it. A distribution having this property is called a *passive* distribution. Furthermore, any collection of systems that are all thermal systems at the same temperature should be also be a thermal system. A distribution having the property such that an array of N copies of the system all have that distribution (and independent of each other) is called a *completely passive* distribution (the terminology is due to Pusz and Woronowicz 1978). It can be shown that the condition of complete passivity

---

[5] The following is indebted to the lucid exposition of Maroney (2007), which synthesizes earlier work.

is satisfied only by canonical distributions.[6] The result extends to quantum systems with infinitely many degrees of freedom, in which case the states singled out are the *KMS states*, which are thus analogs, in that context, of canonical states.

If the Hamiltonian is (at least approximately) a sum of a large number n of component energy terms (as it will be if the system is a collection of subsystems that interact with each other weakly), these terms will, on the canonical distribution, be (at least approximately) probabilistically independent of each other, and the weak law of large numbers applies. The variance in energy, $V(H)$, will be small compared $\langle H \rangle^2$; the ratio of $V(H)$ to $\langle H \rangle^2$ will be of the order of $1/n$. If the system is composed of a macroscopic number of molecules, of the order of 10^{24}, this means that, on the canonical probability distribution, the energy will, with extremely high probability, be *very* close to its expectation value, and fluctuations away from this value will be negligible.

Gibbs also considers another probability distribution, appropriate, not for a system in contact with a heat bath, but for an isolated system. If the system's Hamiltonian is a time-independent function on the phase space, total energy is conserved, and the system's trajectory in phase space is confined to a surface of constant energy, of dimension one less than the dimension of the phase space. We can define an invariant measure on the energy surface as follows. Consider the phase space region between two nearby energy surfaces, separated by an energy difference δE. This will be an invariant set, meaning that phase points within the region will stay within it; furthermore, within this set, Liouville measure will be an invariant measure. For any measurable subset A of one of the surfaces, we can extend it to the region between by considering the set of points that arise from A by displacements along lines perpendicular to the surface, until we reach the second one. Consider the phase space volume of this region on A, in between the two energy surfaces, and take the limit as the energy difference δE is diminished indefinitely. This yields an invariant measure on the energy hypersurface that, following Gibbs's coinage, is called the *microcanonical distribution*.

Gibbs also spoke of *grand ensembles*. These have to do with systems whose total number of molecules is not fixed, as they might be undergoing chemical reactions or exchanging molecules with the environment. A *grand canonical distribution* is meant to be appropriate for a system whose total number of molecules is not fixed and which is in contact with a heat bath. We will not

[6] This is done in the context of quantum mechanics, but it seems that an analog of the argument should go through in the classical context.

need to say anything about these, but I mention them, lest anyone get the impression that statistical mechanics can only be applied to a system with a fixed number of degrees of freedom.

7.3.2 Statistical mechanical analogs of entropy

If some approximation to the laws of thermodynamics is to be recovered, at least for systems possessing many degrees of freedom, then it is necessary to find quantities that behave like thermodynamic quantities.

Entropy is of particular interest. Recall from §6.4 that the concept of entropy requires a distinction between two modes of energy transfer between systems: heat transfer, and doing work. Recall, also, that the Second Law, in any of its statements, also relies on such a distinction. How are we to represent such a distinction, in a mechanical context?

The way that this is standardly done stems from Gibbs (1902). We suppose that the Hamiltonian of a system depends, not only on the canonical variables that characterize the state of the system, but also on certain parameters, regarded as external, manipulable parameters. The values of these parameters are treated as exogenous variables that are set by some process external to the system and whose values are uninfluenced by the vagaries of the system as its phase point wanders about its phase space. As the Hamiltonian is dependent on these parameters, a change in the parameters can effect a change of the system's energy, and energy may be added to the system or extracted from it via such changes. As long as this is the only commerce the system has with the outside world, the evolution is still Hamiltonian, and phase space volume is still conserved—that is, any measurable region of phase space evolves into one with the same measure.

Let these external parameters be $\{a_1, \dots, a_k\}$, and define

$$A_i = -\frac{\partial H}{\partial a_i}. \tag{7.5}$$

These play the role of *generalized forces*, and the work done on the system by a change of one of these parameters from a value $a_i(0)$ to $a_i(\tau)$ is given by

$$W_i = -\int_0^\tau A_i(x(t))\, \dot{a}_i(t)\, dt, \tag{7.6}$$

where $x(t)$ is the system's phase-space trajectory during the process.

Now, suppose we have a system that is initially in thermal contact with a heat bath at some temperature T. The system is disconnected from the heat bath, and then subjected to various operations consisting of small changes to the external parameters, possibly changing its temperature a small amount. At the end of the process, it is placed in contact with a heat bath at the new temperature. We use a canonical probability distribution for initial and final states. These two distributions may be different, as the distribution depends on the temperature, and, via dependence of the Hamiltonian on the external parameters, also on the values on the external parameters.

The energy of the system may, of course, change as a result of the process. Because of uncertainty about the initial state, and about the details of its interaction with the heat baths, we would not attempt to follow the precise trajectory of the system through its phase space and on the basis of that calculate a precise change in energy. However, since we have probability distributions over initial and final states, we can calculate the expectation values of energy in these two states, and hence the expectation value of the change in energy.

It can be shown that the change in expectation value of energy, $d\langle H \rangle$, is given by[7]

$$d\langle H \rangle = -kT\, d\langle \log \rho \rangle - \sum_{i=1}^{k} \langle A_i \rangle da_i. \tag{7.7}$$

Compare this with the thermodynamic relation,

$$dU = TdS - \sum_{i=1}^{k} A_i\, da_i, \tag{7.8}$$

which partitions a change in internal energy of a system into a term TdS denoting heat exchange, and terms denoting the work done on (or by) the system via change in external parameters. About this, Gibbs remarks,

> This equation [our (7.7)], if we neglect the sign of averages, is identical in form with the thermodynamic equation (482) [our 7.8], the modulus (Θ) [our kT] corresponding to temperature, and the index of probability of phase [$\log \rho$] with its sign reversed corresponding to entropy.
>
> (Gibbs 1902: 168)

---

[7] Modulo a change in notation, this is eq. (114) on p. 44 of Gibbs (1902), repeated as eq. (483) on p. 168.

He follows this with,

> We have also shown that the average square of the anomalies on ε [energy of the system], that is, of the deviations of the individual values from the average, is in general of the same order of magnitude as the reciprocal of the number of degrees of freedom, and therefore to human observation the individual values are indistinguishable from the average values when the number of degrees of freedom is very great. In this case also the anomalies of η [$= \log \rho$] are practically insensible. The same is true of the anomalies of the external forces (A_1, etc.), so far as these are the result of the anomalies of energy, so that when these forces are sensibly determined by the energy and the external coördinates, and the number of degrees of freedom is very great, the anomalies of these forces are insensible.

In other words: for macroscopic systems, on a canonical distribution each of the quantities whose expectation values appear in eq. (7.7) will have variance so small as to be negligible, and hence, with very high probability, the actual values will be extremely close to their expectation values, and we can take these expectation values as stand-ins for the actual values, from which they will, with high probability, differ only a little.

It is not obvious from the passage quoted whether it is the "index of probability" (that is, the logarithm of the probability density, or $\log \rho(x)$), which varies with the microstate of the system, or its expectation value $\langle \rho \rangle$, which is a property of the probability distribution, that (when multiplied by Boltzmann's constant k and reversed in sign) is to be taken as analogous to the thermodynamic entropy. The remark about neglecting the sign of averages suggests the former, but, a few pages later, Gibbs takes the expectation value to be the relevant analog of entropy. There is a principled reason why it doesn't matter which it is. Gibbs takes the laws of thermodynamics to be valid only for those macroscopic systems for which the variance of these quantities can be neglected and for which the actual quantity and the expectation value can be treated interchangeably. Despite this qualification, the quantity

$$S_G[\rho] = -k \int \rho(x) \log \rho(x)\, dx \qquad (7.9)$$

has come to be called the *Gibbs entropy*, and is sometimes used without the restriction to systems of many degrees of freedom.[8]

Gibbs also proposed analogs of entropy appropriate to systems with a definite, known energy, represented by a microcanonical distribution. His first proposal is

$$S = k \log \Omega(E), \qquad (7.10)$$

where $\Omega(E)$ is the phase-space volume of the set of states satisfying relevant external constraints and having energy less than or equal to E (Gibbs 1902: 170). Only a few pages later, he considers another analog, namely,

$$S = k \log \omega(E), \qquad (7.11)$$

where $\omega(E)$ is what is called the *structure function*,

$$\omega(E) = \frac{d\Omega}{dE}. \qquad (7.12)$$

An energy shell of small width δE containing the surface of energy E has phase space volume approximately equal to $\omega(E)\,\delta E$.[9]

An argument can be provided that extends the identification of the canonical entropy as the appropriate statistical analog of thermodynamic entropy to systems with arbitrary distributions. The argument goes through in the context of classical mechanics, with the analogy of entropy taken to be the Gibbs entropy, and also in the context of quantum mechanics, with the analog of entropy taken to be the *von Neumann entropy*,

[8] The identification of entropy with S_G, for a canonical ensemble, was arrived at independently by Einstein (1902; 1903). As Uffink (2006) points out, for gases this identification of canonical entropy appears already in Boltzmann (1871b), as eq. (18).

[9] There has been extensive discussion whether the quantity (7.10), called the *volume entropy*, or the quantity (7.11), the *surface entropy*, is the analog of entropy appropriate to an isolated system. For pointers to the discussions, and an argument in favor of the former, see Hilbert et al. (2014).

For those who might be perplexed by all these entropies, it may be helpful to illustrate them with reference to a system for which it is easy to do the calculations, an ideal monatomic gas with N molecules. The thermodynamic state can be characterized by the volume V and temperature T, or equivalently, by volume and internal energy E, which is proportional to the temperature. The thermodynamic entropy difference between two states a and b is equal to

$$S_N[\hat{\rho}] = -k \operatorname{Tr}[\hat{\rho} \log(\hat{\rho})]. \tag{7.17}$$

The argument begins with the identification of canonical distributions as the distributions appropriate to represent thermal systems, that is, systems whose only role is to act as heat sources or sinks. We consider a system A, with which may be associated, at time t_0, an arbitrary probability distribution. We suppose we have available a number of thermal systems $\{B_i\}$, with which our system can exchange heat. The probability distributions for these systems are taken to be canonical. The system may also be manipulated in various ways via changes to its Hamiltonian.

In order to make sense of entropy, we need a work-heat distinction. Changes to the energy of the system A via changes to its Hamiltonian are to be counted as work; changes to its energy via exchanges with the thermal systems B_i are to be counted as heat. Since we are dealing with probability distributions, we do not try to estimate the precise amount of heat transfer, but compute, instead, *expectation values* of the amounts of heat exchanged with the various heat baths. We can invoke a result that, in its classical form, is found in Gibbs (1902: 160–64), and in its quantum version, in Tolman (1938, §128–30).

Suppose that, at time t_0, a system A is uncorrelated with thermal systems $\{B_i\}$, with which are associated canonical distributions at temperatures T_i. At times t_0 and t_1 the Hamiltonian of the joint system contains no interaction terms between A and the thermal systems. Between time t_0 and time t_1 A interacts successively with the thermal systems. Let Q_i be the energy exchanged with thermal system B_i, counted as positive if the energy

$$
\begin{aligned}
S_b - S_a &= kN\left(\frac{3}{2}\log(E_b/E_a) + \log(V_b/V_a)\right) \\
&= kN\left(\frac{3}{2}\log(T_b/T_a) + \log(V_b/V_a)\right).
\end{aligned} \tag{7.13}
$$

If ρ_a and ρ_b are canonical distributions in spatial volumes V_a and V_b, respectively, with temperatures $T_a = 1/k\beta_a$ and $T_b = 1/k\beta_b$, then

$$S_G(\rho_b) - S_G(\rho_a) = kN\left(\frac{3}{2}\log(T_b/T_a) + \log(V_b/V_a)\right), \tag{7.14}$$

in exact agreement with (7.13). For this system, we have,

$$k\log\Omega(E_b, V_b) - k\log\Omega(E_a, V_a) = kN\left(\frac{3}{2}\log(E_b/E_a) + \log(V_b/V_a)\right), \tag{7.15}$$

again, in exact agreement with (7.13), and

$$k\log\omega(E_b, V_b) - k\log\omega(E_a, V_a) = kN\left(\left(\frac{3}{2} + \frac{1}{2N}\right)\log(E_b/E_a) + \log(V_b/V_a)\right), \tag{7.16}$$

which, for large N, closely approximates (7.13).

of A is increased. Then the expectation values of these heat exchanges satisfy,

$$\sum_i \frac{\langle Q_i \rangle}{T_i} \leq S[\rho_A(t_1)] - S[\rho_A(t_0)], \qquad (7.18)$$

where, in the classical context, S is the Gibbs entropy, S_G, and, in the quantum context, the von Neumann entropy S_N.

This is the statistical analog of the Second Law of Thermodynamics (compare eq. 6.13).

We will count a process as *cyclic* if the marginal distribution for A is the same at the end as it was at the beginning. If the system A is initially uncorrelated with the thermal systems B_i, correlations between them may be built up as a result of their interactions, and hence the probability distribution of the joint system may not be the same at the end of the process as at the beginning. However, if we consider only the condition of A and any work that might have been done, then these correlations may be disregarded as irrelevant. If the process is a cyclic one, we have,

$$\sum_i \frac{\langle Q_i \rangle}{T_i} \leq 0. \qquad (7.19)$$

The limiting process, in which

$$\sum_i \frac{\langle Q_i \rangle}{T_i} = S[\rho_A(t_1)] - S[\rho_A(t_0)], \qquad (7.20)$$

corresponds to a thermodynamically reversible process. If there is a reversible process that leads from $\rho_A(t_0)$ to $\rho_A(t_1)$, then this result *uniquely* singles out $S[\rho]$ as the statistical mechanical analog of entropy, if expectation values of heat exchanges are to be considered.

To get a feel for what this means, consider a heat engine operating in a cycle between a hot heat source at temperature T_h and a cooler heat sink at temperature T_c. It extracts an amount of heat Q_h from the hot bath, performs work W, and discards an amount $Q_c = Q_h - W$ into the heat sink. We allow for some uncertainty in the states of the systems, and hence may be uncertain about the exact amounts of these quantities. But equation (7.19) tells us that, as long as the initial probability distribution for the state of the engine is uncorrelated with that of the heat baths, and the operation returns its marginal probability distribution to the original one, then the expectation values of heat extracted and heat discarded satisfy (recalling that a quantity of heat counts as positive if it is going into the engine and negative if it is going out):

$$\frac{\langle Q_h \rangle}{T_h} - \frac{\langle Q_c \rangle}{T_c} \leq 0. \tag{7.21}$$

This gives us, for the expectation value of the work extracted:

$$\langle W \rangle \leq \left(1 - \frac{T_c}{T_h}\right)\langle Q_h \rangle. \tag{7.22}$$

Thus, the thermodynamical bound on the efficiency of an engine operating between these two reservoirs becomes a bound on expectation value of work extracted. The statistical version of the Second Law doesn't prohibit you from getting, on occasion, more work than expected out of a process, but merely says that you can't *consistently and reliably* do it, and thus, in Szilard's analogy, is like a no-go theorem for a gambling system:

> Consider somebody playing a thermodynamical gamble with the help of cyclic processes and with the intention of decreasing the entropy of the heat reservoirs. Nature will deal with him like a well established casino, in which it is possible to make an occasional win but for which no system exists ensuring the gambler a profit.
>
> (Szilard 1972: 73, from Szilard 1925: 757)

Thermodynamics provides us with a definition of entropy differences between equilibrium states. It says nothing about how or whether entropy is to be attributed to states out of equilibrium. However, S_G, as defined by (7.9), and S_N, as defined by (7.17), are well-defined for arbitrary distributions or states, whether or not they are stationary. The question arises, then: can we extend the notion of entropy to systems out of equilibrium?

The canonical distribution has the property that it maximizes $S[\rho]$ among all distributions with the same expectation value of energy, and the micro-canonical distribution maximizes it among distributions possessing the same exact energy. (Gibbs 1902: 144) suggests that, for distributions ρ other than equilibrium distributions, the difference between $S_G[\rho]$ and the maximum value consistent with the relevant external constraints be taken as an indicator of the distance from equilibrium. For a system in thermal contact with a heat bath, it is possible for this quantity to gradually increase towards its maximum. However, for an isolated system, its value is constant in time. If, therefore, the entropy of an isolated system spontaneously increases, then the Gibbs canonical entropy is a poor candidate for a non-equilibrium entropy, and for this reason Gibbs rejects its use as a measure of distance from equilibrium for isolated systems.

To see how to define a notion that *does* behave in the way that one would expect a non-equilibrium entropy to behave, let's ask a seemingly nonsensical question: *does* the entropy of an isolated system increase? Consider again the example of free expansion of a gas. As mentioned above, in §6.5.4, if the gas is isolated, even after it has expanded, you still know a lot more about the state of the gas than would be represented by the microcanonical distribution, and a being with this knowledge and with unlimited powers of manipulation could exploit this knowledge to extract work from the system. Such a being would not regard the expansion as a lost opportunity to do work, and would not regard the expansion as an increase of entropy. For us, it *is* a lost opportunity.

To be useful to you for the purposes of extracting work from the system, the knowledge you have must be relevant to features of the system that you can exploit. Thus, since thermodynamic entropy is relevant to possibilities of extracting work from a system, the probability distributions that are germane to thermodynamic entropy are, not distributions over the full degrees of freedom of the system, but over a limited subset that can be exploited. For such a distribution, S_G *can* increase. Gibbs (1902: 148–50) suggests a coarse-graining of the probability distribution. Partition the phase space into cells that are small but not too small, and, given a probability density ρ, define a new one, $\bar{\rho}$, that, at each point x in the phase space, has the value that is the *average* of ρ over the cell that x belongs to. If the initial distribution converges strongly[10] to an equilibrium distribution, this coarse-grained density function will converge to the equilibrium density function, and $S_G(\bar{\rho})$ will converge towards the equilibrium value.

7.4 The neo-Boltzmannian approach

As Uffink (2017) has noted, the roots of virtually every extant approach to statistical mechanics can be found in the work of Boltzmann, who took an

---

[10] See §8.2.1 for definition.

eclectic approach to the subject. However, the strand of Boltzmann's work highlighted by the Ehrenfests in their well-known survey of the foundations of statistical mechanics (Ehrenfest and Ehrenfest 1912) is what is generally referred to as "Boltzmannian" or "neo-Boltzmannian" statistical mechanics.

7.4.1 Boltzmannian entropy

In this section we follow the procedure of Boltzmann (1877b; 1896), which is summarized in Ehrenfest and Ehrenfest (1912). Suppose the macrostate of the system is defined by the values of a small number of functions $\{X_1, ..., X_k\}$. Partition the accessible phase space Γ into regions corresponding to small intervals of values of these macrovariables; each such region consists of points that, for practical purposes, share the same values of the macrovariables. For any phase point x, let $M(x)$ be the macrostate containing the phase point x.

For example, if the system is a gas consisting of identical molecules, confined to a box, we might partition the volume of the box into small (but not too small) cubes of equal volume, and similarly partition the range of possible momenta of any molecule into small, but not too small three-dimensional cubical regions of momentum space, and take, as macrovariables, a specification, for every pair of spatial cube and momentum cube, of how many molecules there are with position and momentum in those regions.

It can be shown that, for a macroscopic ideal gas, provided that the elements of the phase-space partition are not too small, the macrostate that has overwhelmingly largest phase-space volume is one in which the gas has spatial distribution that is approximately uniform and the frequency distribution of the molecular momenta approximates the Maxwell–Boltzmann distribution, which Boltzmann, following Maxwell, had argued is an equilibrium distribution of momenta of molecules in a gas. This suggests an identification of the largest (in phase-space volume) macrostate as the equilibrium macrostate.

Moreover, it can be shown that, for a macroscopic ideal gas, the logarithm of phase-space volume of this largest macrostate has the same dependence on temperature and volume as does the entropy of an ideal gas. This suggests that take the entropy of any macrostate to be proportional to the logarithm of its phase-space volume. This, in turn, suggests a way of ascribing entropy to phase-points. Let λ be Liouville measure, for cases in which the external constrains don't specify precise energy, or else microcanonical measure, for the case of an isolated system of fixed energy. Any phase-point x will belong to a macrostate $M(x)$, which will have measure $\lambda(M(x))$. Define the *Boltzmann entropy*:

$$S_B(x) = k \log(\lambda(M(x))). \qquad (7.23)$$

This gives an appearance of assigning an entropy to a system that is a property of the physical state of the system alone. But note that the value of the Boltzmann entropy depends, not only on the phase point x, but also on the macrovariables chosen to define macrostates (presumably, these are the ones that we are able to measure), on a partition of the macrovariables into sets fine enough that differences within a set are regarded as negligible (this, presumably, has to do with the precision with which we can measure the macrovariables), and *also* on a particular choice of measure on phase space, λ.

This last point deserves emphasis. Boltzmannian arguments often have an air of involving mere combinatorics, a simple counting of possibilities.[11] But, as always (as Laplace emphasized), this requires a specification of which events are to be regarded as equiprobable. This point can be illustrated with reference to our parabola gadget.

Suppose you have a large number N of parabola gadgets—say, 10,000, to connect with the simulated experiment of §5.5—evolving independently. Define macrostates by partitioning the diagonal into a number of bins b_i of width w_i (in accordance with the example, let's use 20 bins). A macrostate consists of a specification, for each of the bins, of how many gadgets have their ball in that bin, and can be presented in the form of a list $\langle n_1, n_2, \dots, n_{20} \rangle$ that gives, for each bin, its "occupation-number."

A macrostate $\langle n_1, n_2, \dots, n_{20} \rangle$ has state-space volume

$$V(\mathbf{n}) = \frac{N!}{n_1! \times n_2! \times \dots \times n_{20}!} \, w_1^{n_1} \times w_2^{n_2} \times \dots \times w_{20}^{n_{20}} \qquad (7.24)$$

This much is combinatorics. But, to say how much volume a given macrostate occupies, we need also to say how wide the bins are. In our example, we partitioned the diagonal into bins of equal width, as measured by distance along the diagonal. But this is just one measure on the state space

---

[11] And, indeed, that is how Boltzmann presented them: "One could even calculate, from the relative numbers of the different state distributions, their probabilities ..." (Boltzmann 1966b: 192, from Boltzmann (1877a; 1909b: 121).

of the system, and there are others that one might consider. Though the bins correspond to equal intervals of the variable x, they correspond to *un*equal intervals of the variable u, and, as measured by that variable, occupy different proportions of the available space, proportions that are given in the second column of Table 5.1. Other variables will attribute other widths to the bins.

It makes a difference. If we use x as the variable to measure the widths, then the overwhelming majority of the state space of our multi-gadget system is occupied by macrostates on which all the bins are roughly equally occupied, and macrostates that are anything like the one obtained in our simulated experiment occupy only a minuscule volume of the state space. If, however, we use u to measure the widths, then the large macrostates are ones for which the occupation-numbers of the bins are roughly proportional to the u-width of the bins, macrostates, that is, like the one obtained in the experiment.

To put some numbers on it: let us compare the volume of the macrostate obtained in our simulated experiment, presented in the last column of Table 5.1, with the volume of the macrostate with an equal occupation number for each bin. If we use x to evaluate the width of the bins, then the equal-number macrostate is larger than the obtained macrostate by a factor of about 10^{655}. If, however, we use u to evaluate the width of the bins, the situation is reversed. The obtained macrostate is larger than the equal-number macrostate by a factor of about 10^{537}.

The same considerations apply to real-world situations. On the Boltzmannian account, the equilibrium macrostate of an isolated macroscopic system is overwhelmingly large *in microcanonical measure*, but there are other measures on which it is overwhelmingly small.

7.4.2 Entropy, phase-space volume, and probability

There is a temptation to simply identify the probability that a system will be found in a given macrostate with the phase-space volume of the macrostate. Then the equilibrium macrostate becomes the most probable macrostate. There is also a temptation to think that, because there are (on microcanonical measure) more equilibrium states than non-equilibrium states, we can conclude, without further ado, that systems will tend to wander into the equilibrium macrostate and stay there for a long time.

This is, of course, far too quick. It is not enough that the majority of states be equilibrium states; we need to know whether the dynamics of the system take states from non-equilibrium macrostates to equilibrium macrostates.

What is needed is that the majority of states be on trajectories that lead to equilibrium. If we are predicting the behaviour of a system that is relaxing towards equilibrium, something still more is needed. Such a system is in a non-equilibrium macrostate, and was, in its recent past, in a further-from-equilibrium macrostate. If the majority of states in its current macrostate are on trajectories with closer-to-equilibrium states in the near future, then (assuming dynamics invariant under time-reversal), the majority of states in the current macrostate are on trajectories with closer-to-equilibrium states in the recent past—that is, the majority of states in the current macrostate arise via fluctuations from closer-to-equilibrium states. What is needed is that, within the tiny fraction of all the states in the current macrostate that share the system's recent past, the majority of states head towards equilibrium.

Suppose we had all that. Then the question still arises: what is special about the microcanonical distribution? We might be tempted to take this as given *a priori* as a typicality measure, the appropriate measure by which to judge whether this or that behaviour of a system is to be regarded as typical. This runs into an immediate problem: this assumption is at odds with *all* of our experience. On such a measure, macroconditions of thermodynamic equilibrium are typical, yet we find in our surroundings an abundance of systems far from equilibrium. All the empirical evidence we have suggests that, as a typicality measure, microcanonical measure fails miserably, about as badly as one can, as it labels virtually everything we see as incredibly, absurdly atypical.

Of course, one might imagine scenarios on which our experience is misleading. One can imagine scenarios on which what we see is not even close to a fair sample of all that there is, and everything we see is atypical indeed. One such scenario is the Boltzmann–Schuetz cosmology, attributed by Boltzmann (1895) to his "old assistant, Dr. Schuetz," on which the Universe consists of a vast sea of matter whose overall state is thermal equilibrium, with occasional fluctuations here and there away from equilibrium (Boltzmann 1895; 1898: §90). Though they would be mind-bogglingly rare, there would also be low-entropy regions as large as the observable universe. On such a scenario, the states we see around us would not be typical states, as the very existence of living, experiencing beings requires low-entropy matter. One can, without contradiction, maintain that features that are ubiquitous in our experience are rare in the universe.

There is a consequence, however, that Boltzmann seems not to have noticed. On such a scenario, the vast majority of occurrences of a given

non-maximal level of entropy would be near a local entropy minimum, and
so one should regard it as overwhelmingly probable that, even given our
current experience, entropy increases towards the past as well as the future,
and everything that seems to be a record of a lower entropy past is itself
the product of a random fluctuation. Moreover, you should take yourself
to be whatever the minimal physical system is that is capable of supporting
experiences like yours, and your apparent experiences of being surrounded
by an abundance of low-entropy matter to be an illusion. That is, you should
take yourself to be what has been called a "Boltzmann brain."[12]

This is a logically possible scenario. But not only does it involve rejecting
judgments of what is typical that are based on experience (which tells us
that out-of-equilibrium systems are ubiquitous), it even goes so far as to
lead us to reject everything we experience as illusory. Empirical evidence
does not support this cosmology. If one could have any grounds for belief
in such a cosmology, they would have to be purely *a priori* grounds, and
would have to be grounds so strong that they could produce a conviction that
stands in the face of all empirical evidence. Yet it is physics that brought us
to these considerations, physics based on empirical evidence that the world
is to be described, at least approximately, as a large number of molecules
evolving according to Hamiltonian dynamics. A theory that tells us that the
experiments on which it founded are illusory undermines its own empirical
base. Clearly, something has gone wrong!

Another way to cling to equilibrium measures as typicality measures is
to postulate a single, incredibly improbable event at the beginning of the
universe. On this sort of picture, a Big Bang is to be thought of as a process
that selects an initial condition at random, with uniform probability on some
energy surface—like a clumsy God throwing a dart at phase space.[13] I humbly
suggest that, whatever we know about Big Bangs, we have no reason to
think that they operate in such a fashion. There are no grounds whatever
for deeming a far-from-equilibrium early Universe improbable.

---

[12] The term is due to Andreas Albrecht. It first appears in print in Albrecht and Sorbo (2004).
The consequence of the Boltzmann–Schuetz cosmology, that we should take the fluctuation we
are in to be no larger than necessary, seems to have been first pointed out by Arthur Eddington.

> (unless we admit something which is not chance in the architecture of the universe)
> it is practically certain that a universe containing mathematical physicists will at any
> assigned date be in the state of maximum disorganization which is not inconsistent
> with the existence of such creatures.   (Eddington 1931: 452)

The implication one should take oneself to be a brain, rather than an entire mathematical
physicist, was drawn by Martin Rees (1997: 221).

[13] This is the closest I can come to making sense of the frequently-repeated assertion that the
initial state of the Universe was a very improbable one.

The upshot of all this is: whatever judgments may be warranted about probabilities of states of things, they are not to be based on considerations of phase-space volume alone.

7.5 Criticisms of the Gibbsian approach

As mentioned in the introduction to this chapter, neither Gibbs nor Einstein regarded themselves as developing a rival approach to Boltzmann's. Nevertheless, some proponents of certain strands of Boltzmann's work, starting with the Ehrenfests (1912), have disparaged the "Gibbsian" approach.

One target of particular reproach has been the Gibbs entropy S_G. About this, we find Shelly Goldstein writing,

> One of the most important features of the Gibbs entropy is that it is a constant of the motion: Writing ρ_t for the evolution on densities induced by the motion on phase space, we have that $S_G(\rho_t)$ is independent of t; in particular it does not increase to its equilibrium value.
>
> It is frequently asked how this can be compatible with the Second Law. The answer is very simple. The Second Law is concerned with the thermodynamic entropy, and this is given by Boltzmann's entropy (1) [our (7.23)], not by the Gibbs entropy (2) [our (7.9)]. In fact, the Gibbs entropy is not even an entity of the right sort: It is a function of a probability distribution, i.e., of an ensemble of systems, and not a function on phase space, a function of the actual state X of an individual system, the behavior of which the Second Law—and macro-physics in general—is supposed to describe. (Goldstein 2001: 47)
>
> It is widely believed that thermodynamic entropy is a reflection of our ignorance of the precise microscopic state of a macroscopic system, and that if we somehow knew the exact phase point for the system, its entropy would be zero or meaningless. But entropy is a quantity playing a precise role in a formalism governing an aspect of the behavior of macroscopic systems. This behavior is completely determined by the evolution of the detailed microscopic state of these systems, regardless of what any person or any other being happens to know about that state. The widespread acceptance of the preposterous notion that how macroscopic systems behave could be affected merely by what we know about them is simply another instance of the distressing effect that quantum mechanics has had upon the ability of physicists to think clearly about fundamental issues in physics.
>
> (p. 48)

In a similar vein, here's David Albert on the matter of defining an entropy in terms of a probability distribution:

> Let's try to keep our heads on. The sort of entropy we are attempting to get to the *bottom* of here, remember, is the entropy that we ran into in *thermodynamics*. And thermodynamical entropy is patently an attribute of *individual systems*. And attributes of individual systems can patently be nothing other than attributes of their *individual microconditions*.
>
> (Albert 2000: 70)

It seems to me that passages such as these, and similar passages to be found in the writings of others, exaggerate the differences between the Gibbsian and neo-Boltzmannian approaches. For one thing, it is not true that Boltzmann entropy, as defined by (7.23), is determined by a system's microcondition alone. As mentioned above, the definition of Boltzmann entropy requires a choice of macrovariables, a coarse-graining of the macrovariables—fine enough but not too fine—, and a choice of measure, the microcanonical measure, by which to calculate volumes of macrostates. Were there a being with the faculties of perception imagined by Maxwell, it might partition the available phase space so finely that, on its partition, all macrostates would be assigned the same entropy. If there is a natural choice of macrovariables, either this has to do with the means available to agents like us to measure such variables, or else I have no idea what it is that distinguishes these variables from any others. And, if there is a natural choice (or range of choices) of coarse-grainings, again, either this has to do with the precision with which agents like us can measure such variables, or else I have no idea what its significance is.

Not much needs to be said about the objection that, for an isolated system, the Gibbs entropy S_G does not increase. The argument for taking this as an analog of entropy involves a system exchanging energy with a heat bath, and is to be used as a measure of entropy differences between initial and final equilibrium states of such a process. Gibbs does *not* suggest that it be used to extend the notion of entropy to isolated systems out of equilibrium (unsurprisingly, since he is looking for statistical mechanical analogies of thermodynamic quantities, and thermodynamics defines entropy differences only between equilibrium states), and neither has anyone else. Gibbs *does* suggest that a *coarse-grained* version of S_G might be used to track relaxation to equilibrium; this is so much in the spirit of the Boltzmannian approach that it is hard to see an advocate of the neo-Boltzmannian approach objecting to it.

A more central issue is whether entropy is the sort of quantity that reasonably may be taken to depend on a probability distribution, thought of as having to do with epistemic states. As we have seen, the definition of thermodynamic entropy, equation (6.12), relies on a distinction between energy transfer as heat, and energy transfer as work. Moreover, as we saw in §6.2.3, every way there is to state the Second Law of Thermodynamics relies on this distinction.

Maxwell's take on the heat/work distinction is that it is means-relative, and has to do with which features of the system we can keep track of and exploit. Considered in and of itself, without reference to such considerations, all there is is energy transfer, and there is no distinction between the two modes of communication of energy. On this view, it is natural for entropy to be defined relative to an epistemic probability distribution. Those who insist that entropy be a property of the microstate of a system, untainted by epistemic considerations, need to say how it is that *they* construe the distinction between heat transfer and work. Can it be construed in terms of a purely physical distinction between two types of microprocesses?

The standard way to introduce the distinction, which stems from Gibbs and is found today in many textbooks, is in the Maxwellian spirit: one divides the parameters on which the Hamiltonian depends into two subsets: those that can be manipulated and set at will, and all the rest, to which we apply Hamiltonian dynamics. Energy changes due to changes of the manipulable parameters count as work, and the remaining changes count as heat. I don't know of any way to draw the work/heat distinction that doesn't rely on considerations of this flavour. There does not seem to be any way of construing entropy as a matter of nothing more than the microphysical state of the system.

7.6 Statistical mechanics and the Principle of Indifference

Imagine a group of scientists who have been working for some time with the parabola gadgets of Chapter 4. They routinely do their work, not in terms of the variable x indicating the distance of the ball from the left edge of the gadget, but in terms of the variable u, in part because the evolution of the gadget takes a simpler form when written in terms of that variable, in part because the invariant distribution is uniform in that variable. A generation grows up who has never seen the evolution of the gadget expressed in terms of parabolas; they call the gadget the "tent-map" gadget. It comes to be taken for granted that the natural variable to use is u instead of x. They become accustomed to using a distribution that is uniform in u and z, and express this in terms of a "postulate of uniform *a priori* probabilities."

Some then forget that this distribution, and this choice of variable, was chosen because it is picked out by the dynamics as a distribution that other distributions evolve towards, and they start to think of employment of the invariant distribution, the one whose density function is depicted in Figure 4.4, as a straightforward application of the Principle of Indifference. They then proceed to apply it to systems that have not had a chance to equilibrate, and find themselves puzzled by unsatisfactory results of such an application.

Though this parable may seem far-fetched, this is what, in fact, has happened in connection with statistical mechanics.

The standard equilibrium probability distributions, canonical, microcanonical, and the like, were introduced by Gibbs. He argued for their reasonableness on the basis of their being stationary distributions, and hence candidates to represent a state of thermodynamic equilibrium. Tolman, in an influential textbook, introduced what he called the "fundamental hypothesis of equal *a priori* probabilities in the phase space" (Tolman 1938: §23). The phraseology is reminiscent of an invocation of a Principle of Indifference, but it is not a naïve invocation; Tolman was sensitive to the fact that, in order for the hypothesis to have any sense, a choice of measure must be made. The measure invoked is Liouville measure, uniform in canonical variables. To argue for the reasonableness of the hypothesis, Tolman appeals to Liouville's theorem, according to which this measure is invariant under dynamical evolution; this shows that "the principles of mechanics do not themselves include any tendency for phase points to concentrate in particular regions of the phase space" (p. 61). The ultimate justification of the hypothesis, however, is empirical.

> Although we shall endeavour to show the reasonable character of this hypothesis, it must nevertheless be regarded as a postulate which can ultimately be justified only by the correspondence between the conclusions which it permits and the regularities in the behaviour of actual systems which are empirically found.   (Tolman 1938: 59)

By the 1960s, however, we find textbooks that present the postulate as a straightforward application of a Principle of Indifference, with no mention that a choice of variables has been made (see e.g. Reif 1968: 48;

Jackson 1968: 8, 83). There is some awareness that there is something odd about this. Thus, E. Atlee Jackson writes,

> This simple *assumption* [of uniform probabilities] is the basis of all of statistical mechanics. Whether or not it is valid is a matter that can only be settled by comparing the predictions of statistical mechanics with actual experiments. To date there is no evidence that this basic assumption is incorrect. A little thought shows that this agreement is indeed remarkable, for our basic assumption is little more than a reflection of our total ignorance about what is going on in the system.   (Jackson 1968: 83)

It would, indeed, be remarkable if ignorance could yield knowledge; this would be something like magic, akin to spinning straw into gold.

We have seen, in previous chapters, a strong temptation among many writers to take the matter of defining probability to be a simple one of counting possibilities. Though this is widely regarded as an illusion, it seems to me that the illusion still exerts an effect on discussions of statistical mechanics. Some writers make an explicit appeal to the Principle of Indifference. Even when no such appeal is made, something like it seems to be lurking in the background. Consider, for example, the following statements, which are echoed in many writings on the foundations of statistical mechanics.

> Consider now a situation in which there is initially a wall confining a dilute gas of N atoms to the left half of the box V. When the wall is removed at time t_a, the phase space volume available to the system is fantastically enlarged, roughly by a factor of 2^N. (If the system contains 1 mole of gas in a container then the volume ratio of the unconstrained region to the constrained one is of order $10^{10^{20}}$). This region will contain new macrostates with phase space volumes very large compared to the initial phase space volume available to the system. We can then expect (in the absence of any obstruction, such as a hidden conservation law) that as the phase point X evolves under the unconstrained dynamics it will with very high "probability" enter the newly available regions of phase space and thus find itself in a succession of new macrostates M for which $|\Gamma_M|$ is increasing.
>
> (Lebowitz 1999a: 518; 1999b: S347)

According to the Boltzmannian scenario propounded here, the overwhelming majority, as measured in phase-space volume, of phase points in a (very small) initial macrostate of the universe evolve in such a way as to behave—for reasonable times, that are not too large on the time scale defined by the present age of the universe, since the big bang—thermodynamically, with suitably closed subsystems having increasing entropy, exhibiting irreversible behavior, and so on. In other words, typical phase points yield the behavior that it was our (and Boltzmann's) purpose to explain. Thus we should expect such behavior to prevail in our universe.

(Goldstein 2001: 52)

The key to the statistical approach is the idea that, under a plausible way of counting possibilities, almost all the available microstates compatible with the given initial macrostate give rise to future trajectories on which the gas expands. ...in view of the vast numerical imbalance between abnormal and normal states, the behaviour we actually observe is 'typical', and therefore calls for no further explanation.   (Price 2002: 93)

In these quotations, it is taken for granted that phase space volume—that is, a measure that is uniform in canonical phase space variables—is the appropriate way to count possibilities. This, of course, raises the Laplacean question: why should we take this measure, rather than some other measure, as the appropriate way of counting possibilities, the one that should guide our expectations?

As we have seen, there is a real sense in which measures uniform in phase space variables are very *un*natural measures. Such measures overwhelmingly favour equilibrium conditions, and to take seriously the idea that this measure tells us which sorts of conditions are "typical" means taking virtually everything we have ever seen to be ridiculously atypical. If we had been handed these measures by a deity that we regarded as infallible, we might then be justified in declaring the entire Universe to be one monstrously improbable fluke, or in entertaining fantastic conjectures, such as the Boltzmann–Schuetz cosmology, on which everything any sentient observer has ever seen is mind-bogglingly atypical. But we are not in such a situation. The measures were handed to us by Boltzmann and Gibbs, and, as Tolman says, the chief reason we have for taking them to be of physical significance is ultimately empirical evidence. But this empirical evidence stems entirely from circumstances in which some equilibration, even if only partial, has taken place.

As has been repeatedly emphasized in this book, no Prince of Indifference is going to appear to magically define our probabilities for us. Some other avenue of attack is needed. This will be the subject of the next chapter.

7.7 Appendix

In this Appendix we provide proof of the recurrence theorem for classical mechanics.

In the abstract, it applies to repeated application of any map $T : \Gamma \to \Gamma$ to any measure space $\langle \Gamma, \mathcal{S}, \mu \rangle$, such that the measure μ is invariant under T, whether or not Γ is thought of as representing the state space of a physical system, and whether or not T is thought of as representing temporal evolution. We will first present the abstract theorem, and then its application to classical physical systems. Poincaré proved the theorem for a dynamical mapping that is invertible; for such a mapping, as mentioned, an invariant measure is also a conserved measure. Invertibility is not needed, and the theorem we prove holds for any invariant measure, without assuming invertibility of the dynamical flow, so long as the measure of the whole state space is finite.

Theorem 7.1 *Let $\langle \Gamma, \mathcal{S}, \mu \rangle$ be a measure space, with μ invariant under the map $T : \Gamma \to \Gamma$, such that $\mu(\Gamma)$ is finite. For any measurable set A, let B_0 be the set of points in A that are taken out of A by T and never return upon repeated application of T. Then $\mu(B_0) = 0$.*

Proof Let A be any subset of Γ, and let B_0 be, as defined in the statement of the theorem, the set of elements of A that are taken out of A by T, never to return. For any natural number n, let B_n be the set of points that are taken into B_0 by n applications of T.

$$B_n = T^{-n}(B_0).$$

Because of the invariance of μ, each of these has the same measure. That is, for all n, $\mu(B_n) = \mu(B_0)$. Furthermore, the B_ns are all disjoint. To see this, note first that, from the very definition of B_0, for any $n > 0$, $T^n(B_0)$, the set that results from n applications of T to B_0, is disjoint from B_0. Second, we observe that, from the definition of B_n, for any n, $T^n(B_n) = B_0$. Now consider two distinct numbers, r, and $r + s$, with $s > 0$. We have

$$T^{r+s}(B_{r+s}) = B_0;$$
$$T^{r+s}(B_r) = T^s(T^r(B_r)) = T^s(B_0).$$

Now, the two sets, B_0 and $T^s(B_0)$, are disjoint whenever $s > 0$. These are the two sets that result from applying the map T^{r+s} to the sets B_{r+s} and B_r, respectively. Therefore, the sets B_r and B_{r+s} are disjoint, whenever $s > 0$.

Thus, the sequence of sets $\{B_n\}$ is a sequence of sets that are mutually disjoint, and all have the same measure, which is equal to $\mu(B_0)$. For any N, the union of $\{B_n, n = 0, ... N\}$, has measure $N\mu(B_0)$. This cannot be greater than $\mu(\Gamma)$. But, since $N\mu(B_0)$ is not greater than $\mu(\Gamma)$ for any N, no matter how large, it follows that $\mu(B_0) = 0$. $\qquad\square$

Now suppose that we have a temporal evolution: a family of maps $\{T_t\}$ indexed by a temporal index t. For our purposes it does not matter whether these indices take on discrete or continuous values. What does matter is that the family satisfies the condition of being *invariant under time translations*: the absolute time at which one starts considering the evolution does not matter, only the duration of the temporal interval. This is the condition that, for any t, s, for any point x in the state space $T_{t+s}(x) = T_s(T_t(x))$. Then we can consider any time interval τ, and the map T_τ. Time translation invariance means that the evolution though a duration $n\tau$ in length is equivalent to n-fold application of the map T_τ. If we have an invariant measure and a system confined to a region of its state space of finite measure, then the theorem applies.

8
Probabilities in Statistical Mechanics

8.1 Puzzles about statistical mechanical probabilities

How should we think of the probabilities that appear in statistical mechanics? Are they objective chances? Epistemic credences? Something else?

When you think about it, there are some puzzling features of the use of the standard equilibrium probabilities in statistical mechanics. Attention to these puzzles will help us figure out how to think about them.

One puzzling feature is the following. Probabilities are said to be introduced on the basis of our ignorance of the precise state of the system, or our incapacity to treat analytically the equations of motion in order to deliver a detailed account of the evolution of the system, or both. This suggests an epistemic reading. However, we make predictions on the basis of them, that are verified by experiment. This suggests that they be thought of as objective chances. Physicists don't do experiments to learn about their own epistemic states. It looks as if we have to employ *both* the epistemic and the objective concept, in an inconsistent way.

There is another puzzling feature, in the application of the microcanonical measure to isolated systems.[1] Consider a system that is prepared in a far-from-equilibrium state with known energy, and then allowed to evolve freely, in isolation from other systems. Eventually it relaxes to equilibrium, at which time we employ the microcanonical measure to compute probabilities of outcomes of experiments on it. What is the status of this measure? Taking it as an epistemic probability is problematic, as it assigns minuscule probability to something that we in fact know about the system, namely, that it was in a far-from-equilibrium state a short while ago. Taking it as an objective chance distribution is equally problematic. As the system was, indeed, in the far-from-equilibrium state a short while ago, and it evolved

---

[1] Which don't, of course, exist, though they are treated of in textbooks. Think of these considerations as puzzling features of the textbook account of these fictitious systems.

Beyond Chance and Credence. Wayne C. Myrvold, Oxford University Press (2021).
© Wayne C. Myrvold.
DOI: 10.1093/oso/9780198865094.003.0008

without interference in the interim, there is zero chance that it will be in any state that cannot be reached from a state having the attributes of its initial state. But the microcanonical distribution attributes high probability to this set of states, states that have zero objective chance of being the actual state of the system. For an isolated system that starts out far from equilibrium and relaxes to equilibrium, the standard equilibrium distribution, it seems, can't be either an epistemic probability distribution or an objective chance distribution.

8.2 Preliminary considerations on convergence of probability distributions

8.2.1 Two notions of convergence of probability distributions

The second point deserves further comment. There is one sense in which a probability distribution with support in a region of the state space corresponding to a far-from-equilibrium macrostate can, for an isolated system, converge towards the microcanonical distribution, and one sense in which it can't, corresponding to two notions of convergence for probability distributions.

Given two probability functions P, Q, defined on a common algebra of propositions $\mathcal{A}$, we define the *total variation distance* between these two functions as the greatest extent to which the two differ as to the probability of any proposition.

$$\delta(P, Q) = \sup_{A \in \mathcal{A}} |P(A) - Q(A)|. \tag{8.1}$$

$\delta(P, Q)$ is zero only if P and Q agree on the probability of everything, that is, only if they are the same probability function. It takes on its maximum possible value of 1 if there's some proposition that P ascribes probability one to and Q, probability zero.

A sequence $\{P_n\}$ of probability functions *converges in total variation* to a limit probability function P if and only if

$$\delta(P, P_n) \to 0 \text{ as } n \to \infty. \tag{8.2}$$

A sequence $\{P_n\}$ *converges strongly* to P if and only if, for all $A \in \mathcal{A}$,

$$|P_n(A) - P(A)| \to 0 \text{ as } n \to \infty. \tag{8.3}$$

Clearly, convergence in total variation implies strong convergence (and hence, despite the names, convergence in total variation is the stronger notion), but not conversely; strong convergence does not imply convergence in total variation.[2]

Suppose, now, we have, for all $t \geq 0$, dynamical maps T_t on the state space of some system, and define, as usual, the distributions P_t by

$$P_t(A) = P_0(T_t^{-1}(A)). \qquad (8.4)$$

If the maps T_t are invertible—that is, if they never map two distinct points to the same point—then the total variation distance between two probability functions will be a constant of the motion, and P_t cannot converge in total variation to a distribution Q except in the degenerate case in which P_t is always equal to Q. However, strong convergence is possible.

Apply this to our relaxation-to-equilibrium example. Suppose a distribution P_0 initially has support in some region of phase space to which the microcanonical distribution assigns very small probability p. Then, as the system evolves in time, there will at each time be some region of phase space to which P_t assigns probability 1 and to which the microcanonical distribution assigns the small probability p. Which region this is will change with time. In that sense, the distribution P_t is always very different from the microcanonical distribution.

If, however, there is some finite set of measurements that exhausts all the measurements anyone will ever make on the system, corresponding to propositions $\{A_1, \ldots, A_n\}$, then it *is* possible for there to be a time τ after which $\{P_t(A_1), \ldots, P_t(A_n)\}$ are all as close as one would like to their probabilities on the microcanonical measure. After such a time, we can use the microcanonical measure interchangeably with P_t to calculate probabilities for this set of propositions. *Not* because we think that the microcanonical distribution is the correct probability distribution, in any sense of "probability", for the system (there remain propositions that it is radically wrong about), but because its probabilities for the propositions of interest differ only negligibly from those produced by what we *would* regard as the correct probability distribution. In such cases, the microcanonical measure can be used as a

---

[2] If we have a measurable space $\langle \Gamma, \mathcal{A} \rangle$, and a topology on Γ, then we can define another notion of convergence: a sequence $\{P_n\}$ converges weakly to P if, for every bounded continuous function f, the expectation values $\langle f \rangle_{P_n}$ converge to $\langle f \rangle_P$. If we're concerned with measures on $\mathbb{R}^n$, then, for any P that is absolutely continuous with respect to Lebesgue measure, strong convergence and weak convergence coincide. If allow P to be a probability distribution (such as the δ-function) that ascribes finite probability to some set of Lebesgue measure zero, then there can be sequences $\{P_n\}$ that converge weakly to P but not strongly.

surrogate for the unwieldy and complicated distribution that would result from evolving a probability distribution over initial conditions. The usefulness of the microcanonical measure in no way depends on its correctness—and a good thing, too!

Strong convergence of P_t to some equilibrium measure is a sufficient condition for there to be a time τ after which $\{P_t(A_1), \ldots, P_t(A_n)\}$ are all as close to their equilibrium values as one would like. But, since we are concerned only with a fixed limited collection of propositions, this is by no means a *necessary* condition. Strong convergence is the condition that this is true for *every* finite set of propositions. If there is some proposition B about the system such that $P_t(B)$ does not converge to an equilibrium value, this will be of no concern if B is not among the macroscopically accessible features of the system.

8.2.2 A note on ergodic theory

Suppose we have a dynamical system whose state space is a measurable space $\langle \Gamma, \mathcal{S} \rangle$, a dynamical map $T : \Gamma \to \Gamma$, and measure μ that is invariant under T. The system is said to be

- *Strong Mixing* iff, for all $A, B \in \mathcal{S}$

$$\lim_{n \to \infty} \mu(T^{-n}(A) \cap B) = \mu(A)\mu(B).$$

- *Weak Mixing* iff, for all $A, B \in \mathcal{S}$,

$$\lim_{n \to \infty} \frac{1}{n} \sum_{k=0}^{n-1} |\mu(T^{-k}(A) \cap B) - \mu(A)\mu(B)| = 0.$$

- *Ergodic* For every subset $G \in \Gamma$ with $T^{-1}(G) = G$, either $\mu(G) = 0$ or $\mu(G) = 1$. (This is one of several equivalent characterizations of ergodicity.)

These conditions are arranged in order of increasing strength; *Strong Mixing* $\Rightarrow$ *Weak Mixing* $\Rightarrow$ *Ergodic*. They form part of the *ergodic hierarchy* (see Berkovitz et al. 2006; Frigg and Kronz 2020 for more on this).

If a dynamical system is strong mixing, then every probability measure that has a density with respect to μ strongly converges to μ upon repeated application of T (and conversely). There is a vast literature on ergodic theory, and some have taken it to be of integral importance to statistical mechanics.

For our purposes, it may be interesting to know that an evolution is (for example) strong mixing, but knowing this is in one sense more than is required, and in another sense less. More, because strong mixing is the condition that $\rho_t(A)$ converges to $\mu(A)$ for *every* measurable set, whereas we need only concern ourselves with a limited set of macroscopically ascertainable properties of the system. Less, because the mere statement of strong mixing says nothing at all about rates of convergence. Statements about long-term behaviour of a system in the infinite limit tell us *nothing at all* about what might or might not happen in any finite initial segment of the evolution, and, if, say, the relevant limiting behaviour requires several multiples of the Poincaré recurrence time, it is absolutely no interest in physical applications.

On the other hand, it is in many cases possible to extract bounds on rates of convergence from proofs regarding long-term limiting behaviour, and, insofar as this is possible, some of the literature on ergodic theory may be relevant to statistical mechanics.

8.3 Equilibration as convergence of probability distributions

Suppose you have a system subjected to some preparation procedure, that is such that you take there to be *either* some objective-chance distribution C_0 over the possible microstates that could result from the preparation, or else some class $\mathcal{C}_0$ of credence functions that represents the range of reasonable credences an agent could have about initial conditions, given the preparation. Evolve these distributions forward, using the actual dynamics of the system. If these preparation probability distributions over initial conditions converge towards some attractor distribution, and if there is a limited set of variables of interest that we wish to investigate, then, after a time sufficiently long for such convergence to have occurred, the attractor distribution will approximate the probabilities yielded by the time-evolutes of the preparation probabilities. If the preparation probabilities are objective chances, these probabilities for events after equilibration will also be objective chances. If the preparation probabilities are epistemic, the equilibrium probabilities will be epistemic chances. This process of convergence of probability distributions is, I suggest, how we should think of the process of equilibration.

Consider, for example, the parabola gadget. Any probability distribution over initial conditions that has a density with respect to the invariant measure converges strongly to that measure, and density functions that don't vary rapidly over the state space converge quickly.

Also, if we have more than one gadget, whose initial states might be correlated, the process of equilibration will tend to erase those correlations. The final condition will be one on which the states of the gadgets are effectively uncorrelated. For a system consisting of a large number of parabola gadgets, we can be reasonably confident that, after sufficiently many iterations, it will end up in a state in which the number of balls in any not-too-small interval of the diagonal is roughly proportional to the μ-measure of the diagonal. *Roughly*, but not exactly, and we will also be able to calculate the probability of any size of deviation of these occupation-numbers from their expectation values.[3]

8.4 Statistical mechanical probabilities as epistemic chances

Suppose we have a system whose dynamics are such as to lead to equilibration, construed as convergence of probability distributions over initial conditions, said probability distributions to be construed as either objective chance distributions or credences. If the input probabilities are objective chances, so are the output probabilities. If the input probabilities are credences, the equilibrium probabilities are epistemic chances. Of the two options— objective chance, and credence leading to epistemic chance—the latter is compatible with deterministic physics, and hence a candidate for making sense of probabilities in classical statistical mechanics. It will also, I will argue in the next chapter, play a role in quantum statistical mechanics, as the dynamics invoked therein is typically deterministic, Schrödinger evolution, and there is a role to play for uncertainty about the initial quantum state.

Thinking about probabilities in statistical mechanics as epistemic chances resolves our two puzzles about statistical mechanical probabilities. One puzzle was how probabilities introduced on the basis of incomplete knowledge could be used to generate testable empirical predictions. Neither a wholly objective nor a wholly epistemic conception of probabilities could have

---

[3] A note of caution—our system of parabola gadgets consists of a large number of non-interacting subsystems. Similar considerations hold for weakly interacting subsystems. For systems consisting of strongly interacting subsystems, we would not expect the equilibrium probability distribution to be one on which the states of subsystems are uncorrelated.

these features. But epistemic chances, recall, combine physical and epistemic considerations in their definition. Unlike credences, their values are shaped by the dynamics of the system, and, unlike credences, they are things that can take on values that are not known to any agent. These values are things that we can formulate hypotheses about, and we can empirically test those hypotheses in the manner outlined in §5.5.

The second puzzle is also easily resolved by epistemic chances. For an isolated system that is known, at time t_0, to be in a far-from-equilibrium state, nothing in the original credence-set $\mathcal{C}_0$ will evolve into something that matches, or even comes close to in norm-distance, the microcanonical measure. However, for any finite set $\mathcal{A}$ of observables, it is possible, after a finite time, for every member of $\mathcal{C}_0$ to evolve into probability functions that yield probabilities that differ inconsequentially from those yielded by the microcanonical distribution. If we have good reason to believe we are in such a situation, we would be well advised to use the microcanonical distribution as a *surrogate* for time-evolutes of the members of $\mathcal{C}_0$. The latter will be unwieldy and enormously complicated, so complicated, in fact, that it might be beyond our power to even write them down explicitly. If the simple and uncomplicated microcanonical measure yields virtually the same results for all propositions of interest, then you would be well advised to work with it, not because it represents reasonable credences about the full set of propositions about the system, but because its values on the propositions of interest are informative about the probabilities we are actually concerned with, namely, the time-evolutes of reasonable credences about initial conditions.

8.5 On the way to equilibrium: partial equilibration, and the Markovian recipe

8.5.1 Partial equilibration

Consider, again, the case of Brownian motion. As mentioned in Chapter 1, the changes in the velocity V of a Brownian particle suspended in a fluid medium satisfy, to a good approximation, the *Langevin equation*,

$$m\frac{dV}{dt} = -\gamma V + F_{ext} + f, \qquad (8.5)$$

where γ is a constant connected with the viscosity of the medium, F_{ext} represents any external force (such as gravity), and f is the fluctuating force due to collisions with the molecules of the medium. Now, if you think about it, any frictional force on the particle is also due to collisions with the molecules of the medium. If the Brownian particle is moving with respect to the medium, that is, moving with respect to a reference frame in which the mean velocity of the medium's molecules is zero, then the *relative velocity* of impacts will tend to be larger in the forward direction—the particle will, on average, be hit harder from the front than from the back—and so there will tend to be a net retarding force, and it can be shown that this is proportional to the velocity of the particle. What the first term on the right-hand side of this equation does is to extract, from the net force due to impacts, the expectation value of the force. The remaining term is deviations from the expectation value.

The last term, therefore, has expectation value zero. It is standardly assumed to have finite variance. Also, it is assumed that correlations between forces at different times are negligible.[4] That is, if you're a Brownian particle, knowing what just hit you doesn't tell you anything about what's about to hit you next. From these features alone, it follows that, if the external force F_{ext} is zero, the expectation value of the square of the velocity satisfies,

$$\langle V(t)^2 \rangle = \frac{\langle f^2 \rangle}{2m^2\gamma} + \left(\langle V(0)^2 \rangle - \frac{\langle f^2 \rangle}{2m^2\gamma} \right) e^{-2\gamma t}. \tag{8.6}$$

Therefore, $\langle V(t)^2 \rangle$ decays to an equilibrium value,

$$\langle V(t)^2 \rangle_{eq} = \frac{\langle f^2 \rangle}{2m^2\gamma}. \tag{8.7}$$

If, moreover, we assume that the probability distribution for velocity of the Brownian particle at least approximates a canonical distribution, we have,

$$\langle V(t)^2 \rangle_{eq} = \frac{3}{2}kT, \tag{8.8}$$

---

[4] Mathematically, this is represented by having the correlations be strictly zero for distinct times, which is represented by a delta-function distribution. Don't fret about whether this could be true of some fictional, idealized system, as it doesn't have to. What the condition is doing is generating a good approximation to real systems. As Leo Kadanoff puts it, "The delta function arises because kicks are so short in comparison with our observation time that we cannot perceive their duration and structure" (2000: 121).

and so,

$$\frac{\langle f^2 \rangle}{2m^2\gamma} = \frac{3}{2}kT.$$ (8.9)

This is interesting, because it relates two parameters, one of which, γ, is an estimate of the viscosity of the medium, and the other, $\langle f^2 \rangle$, is a measure of the magnitude of the small buffets delivered to the particle. In light of the kinetic theory, it is unsurprising that a relation between the two exists, as both are due to collisions of the particle with the molecules of the medium.

The fundamental assumption behind the Langevin equation is that, even if the Brownian particle has not yet equilibrated, we can treat the surrounding medium as if it is in equilibrium, despite the fact that the Brownian particle is interacting with it and hence having an influence on its state. This is the sort of assumption that Sklar (1993: 206, 217, 290–91) refers to as a "rerandomization posit." Its rationale has to do with a separation of time scales. Some variables will equilibrate quickly, others, less quickly. The process of equilibration in the surrounding medium occurs so quickly, compared to time-frames relevant to the gradual slowing of the Brownian particle, that the medium can be regarded as effectively in equilibrium at all times.

Much of the work in non-equilibrium statistical mechanics relies on singling out some variables as "slow" variables, whose gradual progress towards equilibrium can be tracked, and "fast" variables, which equilibrate so quickly that we are justified in using an equilibrium probability distribution over them at all times.

This can be illustrated by reference to the parabola gadget. The fact that probability distributions for x approach the invariant distribution rapidly has implications for the evolution of the other variable z, which, as was proven in the Appendix of Chapter 4, equilibrates more slowly.

Suppose we were concerned about tracking the evolution of z, and were either uninterested in or had no access to the value of x. This would pose a problem. Since the value of z on the $(n+1)$th iteration depends, not only on z, but also on x, the probability distribution for z at the $(n+1)$th stage depends on the probability distribution for the full state (x, z) at the nth stage.

For *certain kinds* of distributions, however, we can get an autonomous equation that gives probabilities for future values of z in terms of its present value. If the probability distribution for x yields equal probabilities for x being less than 1/2 and for its being greater than 1/2, then, with probability 1/2, z_{n+1} is $z_n/2$, and, with probability 1/2, it is $1 - z_n/2$. This gives an effective dynamics for z that is stochastic in nature.

$$z_{n+1} = \begin{cases} z_n/2, & \text{with probability } 1/2; \\ 1 - z_n/2, & \text{with probability } 1/2. \end{cases} \qquad (8.10)$$

This, in turn, permits us to get an equation that produces a probability distribution for z_{n+1} from a distribution for z_n.

Not all probability distributions for x have this feature. However, as we have seen, sufficiently nice probability distributions for x rapidly approach the invariant distribution, which has the requisite symmetry. For such initial distributions, after a few iterations, the autonomous equation that yields probabilities for z_{n+1} in terms of the value of z_n will be a good approximation to what one would obtain using the probability distribution over the full state space.

As mentioned, what makes this work is separation of time scales. The variable x changes its value rapidly—nearby values of x, in one iteration, increase their separation, whereas, for almost any fixed value of x, nearby values of z approach each other. Values of x are *unstable*, compared to values of z. As a result, evolution tends to smooth out density functions for x and complicate density functions for z. The fact that x is unstable compared to z is reflected in rapid approach towards the invariant probability distribution for x, compared to the rate of approach of z towards its invariant distribution.

This has an analog in real physical systems. Consider, for example, a helium balloon made of rubber. As every child knows who has had such a balloon, the balloon is permeable to helium and to air, and after a while, the balloon will reach an equilibrium state in which the gas inside it is the same pressure and composition as the atmosphere outside. But this happens slowly, and, in the meantime, we can treat the system as one in which there is helium gas at high pressure inside the balloon and the ordinary atmospheric mix of gases outside. If, on a warm sunny day, you take a balloon that has been outside into an air conditioned building, the contents of the balloon will rapidly come to the same temperature as the air in the room, and we will be able, over time scales short on the time scale of leakage of helium from the balloon, to treat the gas inside the balloon and outside as each being in a state of thermal equilibrium, at the same temperature but unequal pressures.

This phenomenon is not uncommon; all is needed is a separation of time scales, in that some variables equilibrate fairly quickly, others, more slowly. Techniques of this sort are pervasive in non-equilibrium statistical mechanics, to obtain an equation governing a probability distribution over a reduced set of variables—the slow variables—typically a set that falls far short of the full set of variables.

For systems for which this sort of local equilibration occurs, we can formulate a recipe for making predictions about macrovariables. Suppose that you have a set $\{F_1, , \ldots, F_n\}$ of macrovariables that you take to define the state of the system. Let ρ be a density function that represents your credence, or some reasonable credence, about the state of the system. This will yield probability distributions for any measurable function on the state space of the system, and this includes the macrovariables $\{F_1, \ldots, F_n\}$. Suppose that you have reason to believe that, for the purposes of predicting the future evolution of the macrostate, details of the microstate other than those relevant to the present distribution of macrovariables are irrelevant. Then, for the purposes of predicting the future macrostate, you can replace ρ with a density function $\bar{\rho}$ that yields the same probabilities for the macrovariables $\{F_1, \ldots F_n\}$, but which is smoothed out over surfaces that agree on the values of the macrovariables (i.e. $\bar{\rho}$ is a function of the macrovariables $\{F_1, \ldots F_n\}$).

The assumption that what ρ says about the current state that is irrelevant to the values of the macrovariables is also irrelevant to future evolution of the macrovariables amounts to the assumption that, for the purposes of predicting future values of the macrovariables, it doesn't matter whether we use ρ or $\bar{\rho}$. When this obtains, this gives us a recipe for prediction.

The Markovian recipe. For the purposes of predicting future values of the macrovariables $\{F_1, \ldots F_n\}$, replace ρ by $\bar{\rho}$, and forward evolve it.

The postulate that this works has been called the *Markovian Postulate* by Oliver Penrose (1970: §I.5).

There is a temptation to take this postulate as a basic principle of statistical mechanics. It must be handled with caution, however—though it provides a good recipe in the normal course of things, there are exceptions.

One circumstance in which the Markovian recipe fails is in spin echo experiments.[5] These experiments involve a large number of nuclear spins which are initially aligned, but which are precessing at rates that differ by slight and individually unpredictable amounts. Take our macroscopic variables to be the values of these spins, within the bounds of experimental

---

[5] The original reference is Hahn (1950). For an accessible exposition, see Hahn (1953). Blatt (1959), Ridderbos and Redhead (1998), Ridderbos (2002), Ainsworth (2005), and Lavis (2004) discuss, from different perspectives, the significance of spin echo experiments for the foundations of statistical mechanics.

precision. After a sufficient interval of time τ, these are distributed among all possible values in an effectively random way, and, as far as the instantaneous macroscopic variables are concerned, the state is indistinguishable from equilibrium. The state of the system contains a memory of the initial alignment, but, since the small differences in precession rates are not among the macrovariables, this memory is effaced in the macrostate.

A pulse is then applied that reverses all of the spins, and sends them precessing in the opposite direction. It turns out that it is possible to do this with sufficient precision that, at time τ later, a close approximation to the original alignment is restored. Hahn (1953: 8) provides a vivid analogy (illustrated on the cover of the November 1953 issue of *Physics Today*): imagine a number of runners who all begin at the same spot on a circular race track. A starting gun is fired, at which time they all start running, at speeds that differ by small random amounts. After a while, they are randomly distributed about the track. But then the gun is fired again, and all the runners simultaneously reverse their directions and proceed to run at the same speed as before. After an interval of time has elapsed equal to the time between the first and second firings of the gun, the runners have all returned to the starting line.

Markovian evolution is sufficiently common that a large proportion of work in non-equilibrium statistical mechanics is devoted to it. To elevate the Markovian Postulate into a general principle would, however, be an exaggeration.

8.5.2 A temporally democratic Markovian recipe?

The Markovian recipe enjoins one to use a smoothed probability distribution for the purpose of predicting future values of macrovariables. *When the recipe works, it is not because the recommended probability distribution represents either reasonable credence or objective chance about the microstate of the system*; typically, use of the probability distribution will throw away considerable information about the past. When it works, it is because that information has become irrelevant to its future evolution, which is why an apparently insane move, that of throwing away part of what you know, is acceptable. The fact that (in many interesting cases) it works towards the future should not be taken as evidence that we ought to expect it to work towards the past—there is nothing puzzling about why throwing away information about the past is a bad idea when setting credences about the past.

Indeed, if one is convinced that it works in the future direction, one should take it to work towards the past only if one has prior credences about the recent past that favour a closer-to-equilibrium past. Suppose that, for some physical system with dynamics that are invariant under time-reversal, you are convinced that, for the purposes of prediction of macrovariables, any reasonable probability distribution can be replaced by one that is as close to uniform as possible, given the current macrostate. This conviction might be based on calculation, on empirical evidence, or on a combination of the two.

Suppose, now, that you run across a system of that type which you have good reason to believe has been evolving in isolation for some time, and find it in some nonequilibrium macrostate (you may think, if you like, of the familiar example of a thermos bottle containing warm water and some ice cubes). What should you predict, and what should you retrodict, about the system?

For prediction, we've already said that you are convinced that taking a distribution that is as close to uniform as possible, given the current microstate, and evolving it forward, is an effective strategy. One might be tempted to conclude that reversible dynamics, together with the effectiveness of the strategy in the forward direction, either suggests or even *requires* you to take it to be an effective strategy in the reverse direction, in which case you should back-evolve the smoothed-out distribution, and conclude that the system was probably closer to equilibrium in the recent past.

This is too quick, and, in fact, is justified only if you take the role of the uniform, invariant distribution to be one that guides your expectations about what is typical and atypical, instead of the role for which you have a rationale, namely, as a *surrogate* for a distribution that represents your belief state about the system. To see this, let us undertake a Bayesian calculation. Let E be the evidence that the system is, at time t_0, in the observed nonequilibrium state. Let H_{non} be the hypothesis that the system was in some nonequilibrium state with the property that application of the recipe of forward-evolving a smoothed distribution over that macrostate accords high probability to E: e.g. a thermos with larger ice cubes and warmer water. Let H_{eq} be a state of the sort that you would expect if you applied the Markovian recipe in the backwards time direction. This will be a state closer to equilibrium than the observed macrostate, for instance, a state with smaller ice cubes and cooler water than in the state you found at t_0. That is, H_{eq} is the hypothesis that, prior to your observation at t_0, some of the water spontaneously froze, giving up heat to the remaining water. Bayes' theorem tells us that

$$\frac{Pr(H_{non}|E)}{Pr(H_{eq}|E)} = \frac{Pr(E|H_{non})}{Pr(E|H_{eq})} \times \frac{Pr(H_{non})}{Pr(H_{eq})}. \tag{8.11}$$

Since H_{eq} and H_{non} are hypotheses about the state at an earlier time than the observed state involved in E, we can apply the recipe of forward-evolving a smoothed probability distribution to compute $Pr(E|H_{non})$ and $Pr(E|H_{eq})$. By construction the former is large; the latter will be tiny. Thus, the first factor on the right-hand side of (8.11) is *huge*. Therefore, $Pr(H_{non}|E)$ will be much, much larger than $Pr(H_{eq}|E)$, and you should regard it as much more likely that the system reached its observed state from a further-from-equilibrium state, *unless* you take the second factor to be minuscule, that is, unless you take the prior probability of H_{eq} to be enormously larger than the prior probability of H_{non}.

That is, if you are convinced, by experience or otherwise, that evolving a smoothed distribution over the system's macrostate works well in the forward direction, you should apply it also in the reverse direction *only if you are already convinced* that your priors about the recent past should be set according to a measure according to which equilibrium states are probable and nonequilibrium states mind-bogglingly improbable. Time-reversibility of the dynamics does not mandate a conclusion that, if a smooth-and-evolve strategy works in the forward direction, it works also in the reverse direction.

This is worth unpacking a bit more, as some readers may be inclined towards a position on which the past-oriented and future-oriented directions of time are treated on an equal basis, and may be tempted to adopt a temporally symmetric version of the Markovian recipe, one that recommends using it for both predictions and retrodictions. The problem with this is that, out of equilibrium, the two-way Markovian recipe could be reliable only at isolated moments of time; one cannot expect it to be reliable at every time during an extended interval during which the macrostate is changing.

To see this, suppose that the forward Markovian recipe works at time t_0, at which time the macrostate is some far-from-equilibrium macrostate M_0. Suppose, also, that applying the forward recipe at time t_0 yields a prediction that, with high probability, at some later time $t_1 = t_0 + \delta$ the macrostate will be in a macrostate M_1, still a non-equilibrium macrostate, but closer to equilibrium than M_0, and that at a still later time $t_2 = t_1 + \delta$, the macrostate is an even closer to equilibrium macrostate, M_2. If all goes well, there will be harmony between applying the forward Markovian recipe at times t_0 and t_1. This will be the case if a smoothed distribution over the macrostate M_1 yields high probability that the macrostate at t_2 is M_2.

However, we would run into trouble if we expected to be able to apply the two-way Markovian recipe at t_1. Assume that—as is usual in statistical mechanics—the macrovariables $\{F_i\}$ used to define the macrostate are invariant under time-reversal (which is the case if they depend only on coordinates and even powers of velocities), and that the dynamics are time-reversal invariant. Then applying the Markovian recipe in reverse at time t_1 yields high probability that the macrostate at t_0 is M_2, and only low probability to the actual macrostate M_0. In the scenario imagined, if the forward Markovian postulate is reliable (in the sense of ascribing high probability to what actually occurs) at t_0, the reverse Markovian postulate is unreliable at t_1.

The same reasoning can be applied in the reverse direction: if the reverse Markovian postulate is reliable at t_0, then the forward Markovian postulate is unreliable at a time $t_1' = t_0 - \delta$ prior to t_0. The two-way recipe, if it is reliable at any time t_0 at which the macrostate is a non-equilibrium macrostate, is unreliable at any time in any interval $t_0 \pm \delta$, where δ is long enough for there to be a change in macrostate but shorter than the time scale of relaxation to equilibrium. The only times at which we would be justified in applying a two-way Markovian postulate would be those special times that are the peak of a fluctuation away from equilibrium, with closer-to-equilibrium states both before and after.

8.5.3 The Markovian postulate and the "generalized H-theorem"

If we have a smoothing, or coarse-graining operation, of the sort discussed in the previous section, and if the evolution satisfies the Markovian postulate, then it is possible to prove an entropy-increase theorem, which Tolman has called (somewhat misleadingly) the *generalized H-theorem*. First, some definitions.

Definition 8.1 *Given a coarse-graining operation, $\rho \rightarrow \bar{\rho}$, an evolution T_t is Markovian for ρ if and only if the result of evolving ρ via T_t, and then coarse-graining, is the same as first coarse-graining, evolving via T_t, and then coarse-graining the result. That is, T_t is Markovian for ρ if and only if*

$$\overline{T_t \bar{\rho}} = \overline{T_t \rho}.$$

Note that whether an evolution T_t is Markovian for ρ depends both on the coarse-graining operation, and on ρ. We will be interested in cases in which the evolution is Markovian for all reasonable ρ for a coarse-graining in terms of macroscopically accessible variables. Note also that, if ρ is itself a coarse-grained distribution—that is, if $\rho = \bar{\rho}$—then, trivially, any evolution is Markovian for ρ.

Let S_G be the Gibbs fine-grained entropy.

$$S_G[\rho] = -k \int \rho(x) \log \rho(x) dx. \tag{8.12}$$

We can also define a macroscopic, or coarse-grained entropy, which uses the smoothed density function $\bar{\rho}$.

$$\bar{S}_G[\rho] = -k \int \bar{\rho}(x) \log \bar{\rho}(x) \, dx. \tag{8.13}$$

We consider the evolution of this macroscopic entropy $\bar{S}_G$, under the assumption of the Markovianity condition. It is easy to show that, under this condition, $\bar{S}_G$ is non-decreasing.

Proposition 8.1 *(Tolman 1938, §51).[6] For an isolated system, if T_t is Markovian for ρ, then*

$$\bar{S}_G[T_t\rho] \geq \bar{S}_G[\rho].$$

The proof depends on two things. One is the fact that the fine-grained Gibbs entropy, $S_G[\rho]$, is conserved under the evolution. The other is that the smoothing operation we've been considering does not decrease S_G.

Lemma 8.1 *For any density function ρ, $\bar{S}_G[\rho] \geq S_G[\rho]$, with equality if and only if $\rho = \bar{\rho}$ almost everywhere.*

This follows from a simple mathematical lemma, which is found in Gibbs (1902, ch. 11, Theorem IV: 132), and Tolman (1938: 171–172).

With this in hand, the Generalized H-Theorem is trivial. We assume Markovianity:

$$\overline{T_t\rho} = \overline{T_t\bar{\rho}}. \tag{8.14}$$

---

[6] Tolman assumes that the starting density is coarse-grained, in which case, as we have already remarked, it is trivially true that T_t is Markovian for ρ. What is stated here is thus a slight generalization of the proposition found in Tolman.

Then we have

$$\bar{S}_G[T_t\rho] = \bar{S}_G[T_t\bar{\rho}] \geq S_G[T_t\bar{\rho}] = S_G[\bar{\rho}] = \bar{S}_G[\rho]. \qquad (8.15)$$

The argument doesn't depend on precise nature of the smoothing operation that takes ρ to $\bar{\rho}$; what matters is that the operation not decrease the Gibbs entropy S_G. Nor does it depend on any other property of S_G other than the fact that it is conserved. The general case is the following.

Lemma 8.2 *Let $\langle \Gamma, \mathcal{A}, \mu \rangle$ be a measure space. Let $\mathcal{P}$ be the set of probability measures on $\langle \Gamma, \mathcal{A} \rangle$. Let T be a measurable mapping of Γ to itself. Let S be a functional on $\mathcal{P}$, and let $P \to \bar{P}$ be an operation from $\mathcal{P}$ into $\mathcal{P}$. Define $\bar{S}$ by $\bar{S}[P] = S[\bar{P}]$. Assume*

(i) $S[T(P)] = S[P]$.
(ii) $S[\bar{P}] \geq S[P]$.
(iii) $\overline{T(P)} = T(\bar{P})$.

Then $\bar{S}[T(P)] \geq \bar{S}[P]$.

Note that the bare mathematical facts don't care whether T is forward evolution or backward evolution, or, indeed, whether it's a temporal evolution at all. The domain of useful applicability of the theorem, however, will be temporally asymmetric. Whenever partial equilibration washes out the past history of a system, for the purposes of prediction, the Markovianity condition will be satisfied for all ρ that represent reasonable credence functions. This will not be the case for backwards-in-time evolution—if there's any possibility that records of the past state of a system exist, then it is *not* the case that the current state of the system is all that matters to credences about its past!

8.6 Explaining thermodynamics

8.6.1 What is to be explained?

Much of the philosophical literature on the relation between statistical mechanics and thermodynamics focuses on the status of the Second Law. Readers with some acquaintance with that literature will reasonably come to the conclusion that it is the Second Law that is somehow mysterious. Such readers might have been taken aback at the casual mention, in §7.3.2, that the Second Law, appropriately modified to take into account that it is a probabilistic law, comes out as a *theorem*, on the basis of assumptions about the physical properties of heat baths, some properties of Hamiltonian dynamics, and nothing else.

To recapitulate: The original version of the Second Law cannot be strictly true. One version of the Second Law says that, in any process that takes a system from a state a to a state b, the heat absorbed into the system must satisfy

$$\int_a^b \frac{dQ}{T} \leq S_b - S_a, \tag{8.16}$$

where S_a and S_b are the entropies associated with the initial and final states. The statistical version concerns itself with probability distributions over microstates, associates Gibbs (in the classical case) or von Neumann (in the quantum case) entropies with distributions, and concerns, not the amount of heat absorbed, but the *expectation value* of heat absorbed. The statistical version of the Second Law doesn't prohibit you from getting, on occasion, more work than expected out of a process, but merely says that you can't *consistently and reliably* do it, and thus, in Szilard's analogy, is like a no-go theorem for a gambling system. The statistical version says that, for any process that takes a system from a state a to a state b (these states being characterized by the values of exogenous variables and by probability distributions over the states of the system), absorbing heat from reservoirs with which we associate canonical distributions and which are taken to be initially probabilistically independent of the system of interest, the expectation values $\langle Q_i \rangle$ of heat exchanged with these reservoirs must satisfy

$$\sum_i \frac{\langle Q_i \rangle}{T_i} \leq S_b - S_a, \tag{8.17}$$

with the limiting case of equality approached arbitrarily closely as reversibility of the process is approached.

The proof of the statistical version of the Second Law assumes that one has available various systems that can be thought of as thermal systems, with which are associated canonical distributions, proven to be the unique distribution appropriate to a system of that sort. These are the heat baths from which heat is absorbed and into which heat is discarded, and the existence of systems like that is necessary to make sense of heat transfer, as distinguished from other modes of energy transfer.

The proof of the statistical version of the Second Law says nothing about where such systems come from; it simply assumes that they exist. The origin of a thermal system is, of course, equilibrium; the thermal systems are those that have relaxed to thermal equilibrium. It is *this* process that remains to be explained.

The First Law of Thermodynamics is just the principle of conservation of energy, and so is a consequence of Hamiltonian dynamics for any time-independent Hamiltonian. The statistical version of the Second Law is a theorem. What remains to be explained is the process of equilibration. There is something to be said for regarding this as outside the domain of thermodynamics (recall our discussion in §6.3). If so, the task of explaining thermodynamics in statistical mechanical terms is not as daunting as the philosophical literature would suggest!

But we do need to explain equilibration. Let's talk a bit about how such an explanation might go.

8.6.2 What counts as an explanation?

A common response to the introduction of epistemic considerations into statistical mechanics is that explanation of the behaviour of physical systems is thereby lost.[7] In particular, showing that the dynamics of physical systems exhibit the right sort of sensitivity to initial conditions required for the emergence of epistemic chances will seem, to some readers, to be in principle irrelevant to anything having to do with thermodynamics.

---

[7] Some samples:

"This interpretation leads to the absurd result that the molecules escape from our bottle because we do not know all about them, and because our ignorance is bound to increase unless our knowledge was perfect to begin with. I believe that this is palpably absurd, and that hot air will continue to escape even if there is nobody in the quad to provide the necessary nescience" (Popper 1982: 109).

"Can anybody seriously think that our merely being *ignorant* of the exact microconditions of thermodynamics systems plays some part in *bringing it about*, in *making it the case*, that (say), *milk dissolves in coffee*? How could that *be*? What can all these guys have been *up* to?" (Albert 2000: 64).

"If the probability appealed to in the explanation is merely a subjective degree of belief then how can it account for the melting of the ice cube? What could your ignorance of the initial state of the gas have to do with an explanation of its behaviour?" (Loewer 2001: 611).

Let us, then ask, how might one explain equilibration? To focus our attention, let us return once more to the parabola gadget, and to the simulated experiment of §5.5. In that experiment, we set 10,000 parabola gadgets going, let them run through 100 cycles of the system's dynamics, and reported a coarse-grained outcome that consisted of dividing the diagonal into 20 bins and reporting, for each bin, how many of the gadgets's balls were located in that bin. As you may recall, the relative frequencies obtained approximated the measures of the bins, on the invariant measure μ (see Figure 5.2).

An explanation is an answer to a why-question, and, as van Fraassen (1977; 1980) has emphasized, why-questions are contrastive; they take (sometimes implicitly) the form: Why this rather than that? Let us, therefore, imagine a hypothetical agent, who is puzzled by the result obtained, because it is nothing like the result he expected. This agent, whom we will call Boltzmannian Bob, has done a combinatorial analysis, and, as a result of his calculation, has concluded that there are vastly many more ways for the balls to be roughly evenly distributed among the bins than there are to achieve the wildly uneven distribution that was obtained—by a factor of about 10^{655}. Bob is, therefore, puzzled by the obtained result.[8] What could one say to alleviate his puzzlement?

Here's what I would do. I would invite him to focus, not on the end result, but on the process leading up to it. What sorts of initial conditions would lead to an equal distribution of balls, after 100 iterations of the gadget's workings? What initial conditions would lead to a distribution like the one obtained?

The diagonal can be partitioned into 2^{100} intervals $\{\Delta_i\}$, each of which gets mapped, after 100 iterations, onto the full diagonal. These intervals are not of equal size (as measured by x); the ones near the edges are tiny compared to the ones in the middle. In each of these intervals, there is a sub-interval that gets mapped onto the left-most bin by 100 iterations of the gadget. For most of the intervals Δ_i (all except a few close to the edges), that sub-interval occupies a proportion (as judged by x) of about 0.144 of the total length of the sub-interval. The upshot of this is that, even if we use x to measure the bins and thereby take them to be of equal length, the overwhelming majority of initial conditions lead to outcomes in which the number of balls that end up in the left-most bin is about 1,440. Similar considerations apply to the other

[8] He has, of course, tacitly assumed that the bins are of equal width—i.e. that x is the appropriate measure of the width of the bins—but wouldn't you? Like Boltzmann, he has not noticed that this is a substantive assumption.

bins. On Bob's way of counting things, the overwhelming majority of initial conditions lead to a distribution of balls in which the proportion of balls in a given bin is approximately equal to the measure of the bin, as judged by the measure μ.

Consider, now, the two statements, which might be offered as an explanation of the distribution obtained in our experiment.

I. The vast majority of states are ones in which occupation-numbers of the bins are roughly proportional to the μ-measures of the bins.

II. The vast majority of states are ones that get mapped, after 100 iterations, into states in which occupation-numbers of the bins are roughly proportional to the μ-measures of the bins.

There are two important differences between these two statements, both of which are relevant to their suitability to be invoked in an explanation of the final distribution. One is that the first is static. No dynamical considerations are invoked. For that reason, it is completely mysterious how it could explain *what happens*. The second, on the other hand, invokes considerations involving the process that leads up to the state of affairs to be explained. The other difference is that the first is more highly sensitive to the choice of measure invoked in making judgments about the majority of states. The second comes out true whether one uses x or u to measure the width of the bins. However (as we saw in §7.4.1), the first, though true if the variable u is used to measure the width of the bins, comes out radically false if x is used instead.

So, statement II comes closer to what would be required for an explanation of the result obtained. But it's still not enough, as it still makes no connection with *what we should expect*.

Any explanation of the observed result must involve an argument that Bob should regard it as improbable that the initial conditions be focused on the minuscule subset of conditions that lead to distributions of balls that are nothing like the one obtained. A key point is not only the smallness of the set, but also its fragmentary nature: for any initial conditions that lead to an abnormal distribution of balls, small perturbations will produce initial conditions that lead to distributions like the one obtained, provided that these perturbations are indifferent as to outcome, that is, provided that the probability of a perturbation is independent of the effect it has on the final outcome of the experiment. If Bob regards it as highly likely that whatever process it is that leads up to the setting of initial conditions, it is not one that preferentially targets initial conditions that lead to abnormal distributions—a fairly weak requirement, given the daunting nature of the task—then Bob's credences over initial conditions must be of the sort that we proved, in Chapter 4, would closely approximate the invariant distribution μ with regards to states of affairs 100 iterations down the line.

These considerations should convince Bob that a result of the sort obtained *is the sort that he should expect*, and this should alleviate his puzzlement. They key point is the absence of any process setting initial conditions that can *consistently and reliably* produce initial conditions that lead to abnormal results.

There's a regress here, but not a vicious one. To get Bob to revise his credences about what the outcome of the experiment should be, we invited him to consider the process that leads to that outcome, and to consider what reasonable credences about initial conditions of the experiment would be. And that involved considerations of the process that led up to the setting of those initial conditions. It might seem that we need to push this all the way back to Bob's credences about the early Universe, which might be fuzzy and wooly indeed. But we don't. The conditions on Bob's credences about initial conditions of our bounded-in-time-and-space experiment required to argue that Bob should expect an outcome like the one we obtained are sufficiently weak that their satisfaction will be largely insensitive to his credences about other matters, in particular, to any credences he might have about cosmological matters. Again, all is needed is that he regard it as highly improbable that the initial conditions of our experiment are the sort that can consistently and reliably lead to abnormal results. This is enough to get him to understand why a result of the sort obtained is what he should have expected. This is one sense of the word "explanation," and the sense relevant to alleviating Bob's puzzlement.

In a similar vein, Roman Frigg (2010: 91) has suggested that the sense of explanation relevant to explaining equilibration is *rational expectability*. "On this conception of explanation we explain X by showing that it is rationally expectable that X occurs." This is exactly what is provided by an explanation in terms of epistemic chances, as such an explanation runs: the dynamics of the system are such that the credences of any reasonable agent, when evolved forward, yield high probability for a result similar to the one obtained. Jos Uffink puts the point nicely,

if we use subjective probability in statistical physics, it will represent our beliefs about what is the case, or expectations about what will be the case.

And the results of such considerations may very well be that we ought to expect gases to disperse, ice cubes to melt, or coffee and milk to mix. They do not cause these events, but they do explain why or when it is reasonable to expect them.   (Uffink 2011: 45)

To answer skepticism of the sort voiced in the quotations in footnote 7, above: of course, our ignorance of initial conditions plays no role in bringing about the mixing of milk and coffee. The dynamics of the system, together with the initial conditions, brings it about. But the extent of our knowledge about initial conditions *is* relevant to what we should expect to happen.

And, of course, our ignorance about the initial conditions is tied to the absence of a mechanism in operation that could reliably put the milk-coffee system (which must be thought of as including everything that could be an influence on it) into a state in which stirring without mixing is going on. If there were such a device, and if it were in operation, it would be reasonable for me to expect non-mixing.

8.6.3 Temporal asymmetry

As mentioned in §6.5.1, it was the reversibility argument that convinced the founders of statistical mechanics—first the British physicists, then Gibbs, then Boltzmann—that thermodynamic behaviour could not be explained on the basis of mechanical considerations alone, and that considerations of probability had to be brought in.

There is a line of thinking that, if correct, could be taken to indicate that probabilistic considerations could not possibly be of any avail in reaching conclusions that are asymmetric under temporal reversal. That line of thinking goes like this. There is a one-one mapping of phase space to itself, which consists of leaving all positions and reversing the directions of all velocities. Liouville measure is conserved under this mapping. Therefore, the probability assigned to any behaviour of a system and to its temporal inverse is the same.

The reason that this reasoning does not constitute a dead-end for attempts to explain temporal asymmetry is that Liouville measure is not singled out as having any particular relation to probability (in any of the senses of the word) *except under conditions of thermal equilibrium*, an alternate characterization of which might be that *it is the only situation in which it is uncontroversially true that any sequence of events is as likely as its temporal inverse*. Out of equilibrium, there is no reason, on any conception of probability, to take probability to be conserved under the operation of velocity reversal.

To be a bit more formal: Let T be the velocity-reversal operation, that takes a point x in phase space and maps it to another point x^T with positions unchanged and velocities reversed. There is a corresponding operation that takes a set A of points in phase space and yields another set A^T that is the result of applying the T-operation to all points in A. We can define a corresponding operation on probability distributions. Given a probability function P on the Borel sets of phase space, define another, P^T, by

$$P^T(A) = P(A^T). \tag{8.18}$$

If we have an explanation of the sort discussed in the previous subsection, which invokes some class $\mathcal{C}$ of reasonable credence functions, then, *if the class $\mathcal{C}$ is invariant under the T-operation*, the explanation will work equally well in the forwards and backwards directions. But, once we have abandoned the idea (which was always a mistake, and for which there never was even a shadow of a ghost of a rationale) that equilibrium distributions such as the canonical and microcanonical distributions are appropriate for guiding our expectations in situations away from equilibrium, it is clear that there is no reason to think that applying the T-operation to a reasonable credence function always yields a reasonable credence function.

So, there is no *prima facie* road-block to providing an explanation of temporally asymmetric phenomena in terms that invoke probabilities. We now need to ask what temporal asymmetries there are to be explained.

Perhaps surprisingly, the statistical version of the Second Law is *not* temporally asymmetric. In fact, the statistical version can be stated without reference to the temporal order of the states a and b. If there is a process that connects states a and b, and if the heat reservoirs are uncorrelated with the system of interest at time t_a, then the expectation values of heat transfers into the system in the interval between t_a and t_b must satisfy

$$\sum_i \frac{\langle Q_i \rangle}{T_i} \leq S_b - S_a. \tag{8.19}$$

This is a theorem. It is true whether t_a is earlier or later than t_b, but its useful application will be limited to situations in which t_a is earlier than t_b. We do think that it is possible, and indeed easy, to prepare systems in such a way that their states are effectively uncorrelated with the states of heat baths they will encounter in the future, but we *don't* think it's so easy to prepare systems so that their states are correlated with states of heat baths but Hamiltonian evolution will undo those correlations.

The temporal asymmetry that remains to be explained lies in the process of equilibration. We want to show that the conditions for equilibration hold under fairly weak assumptions about probability distributions over initial conditions, and under realistic assumptions about the dynamics of the systems that equilibrate. This should yield also the conclusion that the states of thermal systems are effectively uncorrelated with those of other systems. Here, "effectively uncorrelated" means that any correlations there might be between details of the microstates that are inaccessible to observation and manipulation are irrelevant to future evolution of the macrostates of things.

We need not fear that the success of such an endeavour will necessarily yield the disastrous conclusion that we should expect things to be closer to equilibrium in the past than in the present. To see why, consider the sorts of credences that you judge to be reasonable ones about the state of a gas.[9]

Let two chambers, one initially empty and the other filled with a gas, be connected by a small hole that permits gas to slowly diffuse from one chamber into the other, and let the entire system be isolated from any influences.[10] Suppose that the diffusion is so slow that, at any time, the gas in each chamber can be treated as effectively in equilibrium at its current pressure. Consider some time τ after the hole has first opened, at which time partial equilibration has occurred; one chamber has lower pressure than the other, but each looks like a gas in equilibrium at its own pressure. The state of the gas in the lower-density chamber (or the state of the gas plus everything that could influence it) is sure to be a state such that, if the velocities were all reversed, would send the gas, after a time interval τ, back into the tiny hole from whence it came. This is entirely to be expected, and there is a perfectly good explanation for why the state has this feature—it's because that's where the gas came from.

Now, consider another situation, in which gas is introduced into the two chambers, at different pressures, so as to reproduce the instantaneous macroscopic state at time τ in the previous example. A small hole is opened

---

[9] When reading the following paragraphs, please use your *actual* judgments of reasonableness, not fictitious ones that your metaphysical scruples would require you to have, if only you could get into that state of mind.

[10] If you're worried that this is impossible, you're right, but then expand the system so that it is large enough to include everything that has an influence on our box of gas, and consider it as a subsystem of that larger system.

up between the chambers. You would, rightly, regard it as highly improbable that the state of the gas (or, worse, the state of it plus everything that is causally relevant to it) be such that all the molecules would soon head for that tiny hole. Nor would it help if someone told you that it *is*, indeed, in such a state, and gave as the reason the temporal inverse of the explanation just given—namely, that the molecules are headed towards the hole because that's where they're going to go!

The notion of causation is inherently time asymmetric, as is the notion of causal explanation. The difference between the sorts of states we judge to be probable, and the sorts of states that would lead to grossly anti-thermodynamical behaviour, is that the latter contain features that, in the absence of a common cause, would seem like unlikely coincidences, coincidences such as all the molecules heading for a hole, with no explanation in their past for this. The probability distributions we deem reasonable reflect this; they contain no correlations between the states of things that are not attributable to events in their common past.[11]

I am undecided as to whether this temporally asymmetric notion of cause should be taken as something fundamental and primitive, something that it makes sense to talk of as operative in the physical world, with no reference to agents who might be interested in doing some explaining, or whether it is something that is only relevant to the doings and knowings and explainings of beings like us. What I *am* sure of, is that, *if* there is no room for talk of causes at the fundamental level, then neither is there any room for talk of explanation at that level. Someone who eschewed all talk of causes at the level of fundamental physics should also be willing to bite the bullet and say that, at that level, there is only the evolving of the world according to its dynamical laws, with no why-relations between events.

8.7 A comment on the "Past Hypothesis"

In his book *Time and Chance*, David Albert introduces, as one of his precepts for reasoning about the world in light of statistical mechanics, what he calls the *Past Hypothesis*, which is the hypothesis

---

[11] This remains true in the face of quantum entanglement. The entangled systems on which we do experiments owe their entanglement to events in the past.

that the world first came into being in whatever particular low-entropy highly condensed big-bang sort of macrocondition it is that the normal inferential procedures of cosmology will eventually present to us.

(Albert 2000: 96)

We are enjoined to take this hypothesis as a constraint on our probabilistic reasonings about things, and to adopt a probability measure that is uniform, in Liouville measure, over the regions of phase space that are compatible with the observed macroscopic states of things and the Past Hypothesis.

Adopting the Past Hypothesis is an instance of a more general prescription, which we may formulate as,[12]

> For any respectable branch of science X, you should have high credence in whatever conclusions that the normal inferential procedures of science, applied to the domain of X, will eventually present to us.

Any readers who are unfamiliar with Albert's book, and are unaware of the context in which the Past Hypothesis is introduced, may well wonder why a hypothesis of this form is deemed necessary. The advice to heed the conclusions of cosmology, or any other respectable branch of science, is good advice, but why does it not go without saying?

The point of introducing the Past Hypothesis, for Albert, is to patch up a faulty statistical postulate. Albert initially introduces a statistical postulate to the effect that one should adopt a probability measure over the set of states compatible with a system's macrostate that is uniform in phase space variables. This, if applied to ourselves and everything around us, would lead, as we have already emphasized, to disastrous retrodictions. This flawed statistical postulate yields high probability that our memories and everything we take to be records of the past are illusions, and that we ourselves are recent fluctuations away from a state closer to equilibrium.

---

[12] The Past Hypothesis, as formulated, carries a presumption that the normal inferential procedures of cosmology will eventually conclude that the world came into being, and that the early universe was a low-entropy highly condensed big-bang sort of macrocondition. Some of this is questionable. It might be the case that cosmology eventually concludes that there was a 'big bounce' instead of an initial singularity. It is also questionable whether there is a well-defined sense in which the early Universe had low entropy (Earman 2006). These are quibbles about the wording of the Hypothesis. Even if the Big Bang followed a Big Bounce, with events prior to it, this Bounce could be taken to be the beginning of our era, and it seems to be well established that, at least locally (that is, within the observable universe), current conditions can be traced back to a hot high-density state. And, if one is worried about the application of an appropriate concept of entropy in this context, the phrase "low-entropy" can be dropped.

It is perhaps worth re-emphasizing—it is the *ordinary* inferential procedures of science that Albert's Past Hypothesis enjoins us to respect. Unless these are first undermined by a faulty statistical postulate, no special postulate is needed to underwrite them. In particular, no special hypothesis about the past is needed to underwrite the usual inferential procedures that take what appear to be records of past events as genuine records.

Suppose, for example, we see what appears to be a somewhat faded photograph of a group of people cheering, with a date "May 8, 1945" pencilled on the back. Call this evidence E. Consider, now, two hypotheses about the state, in 1945, of the matter that you hold in your hands, which now has the appearance of a photograph. Let H_1 be the hypothesis that, in 1945, it had the appearance of a newly printed photograph of the same scene, and let H_2 be the hypothesis that it was a pile of undifferentiated dust. The current appearance of this matter is, on both hypotheses, *possible*. On the supposition of H_1, its current appearance is not unlikely, and, indeed, is the sort of thing one would expect if the hypothesis is true. On the supposition of H_2, its current appearance is the result of a chance fluctuation that, on anything like the usual statistical-mechanical treatment, is enormously, mind-bogglingly, improbable.

The Bayesian analysis invoked in §8.5.2 applies. If Cr is a reasonable credence, $Cr(E|H_1)$ is enormously larger than $Cr(E|H_2)$. The evidence thus strongly favours H_1 over H_2. Posterior credences for H_1 and H_2, in light of the evidence E, satisfy

$$\frac{Cr(H_1|E)}{Cr(H_2|E)} = \frac{Cr(E|H_1)}{Cr(E|H_2)} \times \frac{Cr(H_1)}{Cr(H_2)}. \tag{8.20}$$

As long as one agrees with what was said about the likelihoods—that is, as long as one agrees that what is observed is overwhelmingly more probable on the supposition of H_1 than on the supposition of H_2—then one need not presuppose *a priori* that something like H_1 is true of the past. If one's prior credences in H_1 and H_2 were equal—or even if one's prior credence in H_2 were a paltry 10^{100} times higher than one's prior credence in H_1, the likelihoods swamp the priors, and posterior credence in H_1 is overwhelmingly larger than posterior credence in H_2.

Fortunately, there was never any reason to introduce the faulty statistical postulate in the first place. What one knows about the state of the system involves details about its past that are represented by an intricate structure of reasonable probability distributions over the state of the system. It so happens that these details are usually irrelevant to prediction of future behaviour, and so, for the purposes of prediction, one can replace a complicated probability distribution by a smoothed distribution that erases those details, and use this as a surrogate for one's actual credences, for the purposes of calculation. But this replacement is warranted only if it does no harm!

There is not, and there never has been, any rationale for the faulty statistical postulate. Any attraction it might seem to have, and any temptation one might have to adopt it, seems to stem from the lingering influence of the Principle of Indifference. And, indeed, this is acknowledged by Albert.

> I've been talking about the postulate about statistics up to now as if it more or less amounted to a stipulation that what you ought to suppose, for the purposes of predicting a system's future behavior, if you are given only the information that the system initially satisfies X, is that the systems is as likely to be in any one of the microconditions compatible with X at the initial time in question as it is to be in any *other* one of the microconditions compatible with X at the initial time in question. That's more or less what the postulate amounts to (I think) in the imaginations of most physicists. And that (to be sure) has a supremely innocent ring to it. It sounds very much—when you first hear it—as if it is instructing you to do nothing more than attend very carefully to *what you mean*, to *what you are saying*, when you say that all you know of the system at the time in question is X.
>
> (2000: 62)

But of course, as Albert himself emphasizes (p. 63), "This is all wrong"—a point that has been repeatedly emphasized in this book.

If we could apply probabilistic considerations to the universe as a whole (and this would require constructing a measurable space that includes the dynamical curvature of spacetime in its degrees of freedom), then, whenever the Markovian recipe works, then Albert's fixed-up postulate yields reasonable probabilities. Of course, as Albert acknowledges (p. 67), his postulate is not the only one that will do the trick (this is an important point—what makes statistical mechanics work is that there is a wide range of probability distributions that yield approximately the same probabilities for future values of macroscopically observable quantities). My complaint about Albert's approach is not so much about where he ends up, as the route by which he gets there. He begins by introducing a disastrous statistical postulate for which there is no motivation except the lingering influence of the Principle of Indifference, a statistical postulate that, if accepted, would lead you to the conclusion that you are a Boltzmann brain and that the author of this book does not exist. We are to be rescued from this disaster by postulating something that we never had reason to doubt, namely, that we may rely on the perfectly ordinary inferential procedures of science, when applied within one particular branch of science, cosmology. Instead of taking a leap into a hole of one's own digging and then clawing one's way out, it would, it seems, be better not to take the leap in the first place.

A Past Hypothesis of this sort, which makes no specific proposal about what the state of the early universe, and is introduced as a fix for a faulty statistical hypothesis, must be distinguished from proposals made by physicists that are motivated by the idea that what we have learned about the early universe is in need of explanation. One proposal in this vein is the Weyl Curvature Hypothesis, a constraint on the early universe proposed by Roger Penrose (1979; see also Penrose 1989a,b) as an explanation for what he regards as the puzzling smoothness of the early universe, as evidenced in the cosmic microwave background. This is not a species of the same genus as Albert's Past Hypothesis, as quoted above. The Albert hypothesis is introduced in order to save retrodiction from the threat posed by the faulty statistical hypothesis, so that we may believe what cosmology is telling us about the early universe; the Penrose hypothesis takes as its *explanandum* what cosmology tells us about the early universe, and seeks a lawlike constraint that would render that early state less puzzling.

9

Probabilities in Quantum Mechanics

9.1 Introduction

Much of the discussion to this point has proceeded as if the world conforms to classical physics. It does not. The question naturally arises, therefore, as to what, if any, of the foregoing is applicable to a world such as ours.

On the face of it, probabilities play a very different role in quantum theory than they do in classical physics. Unlike classical physics, quantum theory as typically formulated invokes an explicitly probabilistic postulate, the Born rule, which tells us how to use a quantum state description of a system to calculate probabilities of outcomes of experiments performed on the system.

Now, of course, one might take these probabilities as playing much the same role as they do in classical physics. Einstein himself suggested that quantum state ascriptions might be thought of as analogous to the probability distributions invoked in classical statistical mechanics (Einstein 1936: 340; 1954: 316–17). On this view, a complete microphysical description of a physical system would make no mention of quantum states; these would be associated with preparation procedures and, like the probability distributions of classical statistical mechanics, reflect less than complete knowledge of the state of a system. This motivates a project of attempting to construct a theory on which quantum states play that role. But that project, though well-motivated, does not succeed; there are very good reasons to believe that quantum states represent something in the furniture of the world (see Myrvold 2020b for an argument to that effect).

Quantum states, thus, are not like classical statistical mechanical distributions. Furthermore, it is frequently said that the fundamental revision in world-view that is required by acceptance of quantum theory is a renunciation of determinism. If this is right, it might be thought that the need for a notion that plays a role akin to that of objective chance and which is compatible with deterministic physics evaporates.

Things are not so simple! For one thing, there exists a theory, alternatively known as the de Broglie–Bohm pilot wave theory or Bohmian mechanics,

Beyond Chance and Credence. Wayne C. Myrvold, Oxford University Press (2021).
© Wayne C. Myrvold.
DOI: 10.1093/oso/9780198865094.003.0009

that is completely deterministic and (provided that probabilistic statements are attached to it in an appropriate way) reproduces all the probabilistic predictions of ordinary non-relativistic quantum mechanics. In order for the theory to function, the probabilities invoked in it cannot be mere credences, as, unlike credences but like objective chances, they can be imperfectly known and hypotheses about their values can be subjected to empirical test. If there is no way to make sense of this in the context of deterministic physics, the de Broglie–Bohm pilot wave theory is nonsense. This, of course, will not be my conclusion; I will argue in §9.4 that the probabilities in a theory of this sort ought to be thought of as epistemic chances.

Another reason that things are not so simple is that quantum mechanics, as it is invoked in quantum statistical mechanics, *is* a deterministic theory. Quantum theory, in its standard textbook presentation, has the peculiarity that it has two rules for evolving quantum states. In ordinary circumstances, the state is to be evolved via a deterministic law of evolution, expressed by the Schrödinger equation. When an experiment is done, the state is to be replaced by a new one, corresponding to the experimental outcome, and it is there, and only there, that probabilities enter, in the standard formulation. Quantum statistical mechanics—including, crucially, studies of equilibration—invokes only deterministic, Schrödinger evolution. There can, therefore, be no hope of deriving a monotonic approach to equilibrium from quantum mechanics alone, for the same reasons as in classical physics: reversibility and Poincaré recurrence. As in classical physics, there exist states that will evolve away from, rather than towards, equilibrium. We should not regard it as impossible that the systems we study in fact, possess those states, but merely improbable. And this means we will have to make sense of probability talk of that sort.

There is no consensus as to how (or, indeed, whether) to understand quantum theory as a description of the world. This gives rise to the literature on what is often called the "interpretation of quantum mechanics" (a potentially misleading phrase, because it might suggest that we have a bare uninterpreted formal theory, which we are to endow with a physical interpretation, and because different avenues of approach to the problem may involve distinct physical theories). Avenues of approach to this issue may be broadly classified by their approach to the so-called *measurement problem*. Probabilities play different roles on each of these classes of approach. I will argue, however, the considerations of the sort invoked in previous chapters, namely, evolution of reasonable credences via deterministic physical laws, have a role to play on each class of approach.

In this chapter, we will begin with some remarks about the process of equilibration, as it is studied in quantum statistical mechanics, and then dive into the measurement problem and the role of probabilities in various approaches to it.

For those unfamiliar with quantum theory and the philosophical issues associated with it, I present, in an appendix to this chapter, a brief overview of the formalism of quantum theory, and some of the philosophical issues. This will not be as self-contained as the introductions to probability, thermodynamics, and statistical mechanics. As a supplement, I suggest a pair of *Stanford Encyclopedia of Philosophy* entries: the entry "Quantum Mechanics" (Ismael 2015) for the basic formalism, and the entry "Philosophical Issues in Quantum Theory" (Myrvold 2018b), for an introduction to the basic philosophical issues, and pointers to further references.

9.2 Quantum equilibration

There are two approaches, at first glance strikingly different, towards the study of equilibration. On one approach, one considers an isolated system, but focusses attention on a limited set of dynamical variables of the system, typically thought of as its macrovariables. The other considers a nonisolated system, in interaction with its environment, and tracks the evolution of the state of the system. The two approaches are not as different from each other as might seem at first glance. In each case, we are investigating the evolution of a limited set of degrees of freedom of a larger system and disregarding the rest. The larger system is itself treated as isolated, and hence undergoing Hamiltonian evolution.

Suppose we have a system that consists of one subsystem S, the system of interest, and its environment, E. Suppose the joint system SE starts out in a state in which both S and E are in pure states—i.e. the initial state is a product state. If there are interactions between S and E, the joint system will typically evolve into a state in which the two are entangled. The reduced state of S will then be a mixed state—what is called an "improper mixture," to distinguish it from a proper mixture, which represents a situation in which the system has some pure state, and we are uncertain which state it is. But, as long as attention is focussed on the system S, and not on joint properties of the system and environment, then the improper mixture is operationally and observationally indistinguishable from a proper mixture. The upshot is that, if thermal states are to be represented by mixed states, then these need not represent any sort of subjective uncertainty about the quantum state of the whole joint system SE.

Suppose we have an isolated quantum system that begins in some pure state, and evolves undisturbed. We focus our attention on some limited set of observables of the system, perhaps those that pertain to some subsystem. If the system has a discrete set of energy levels, then the quantum recurrence theorem (see Appendix to this chapter) applies, and the quantum state of the system, and hence also any subsystem, will be uniformly quasi-periodic. That is, after some time τ, which doesn't depend on the initial state, the system will return to a close approximation to its initial state.

In spite of this, it is possible for the state of a subsystem to approach some sort of equilibrium state, and remain there for an enormously long time, much longer than the time period of interest in which we will be interacting with it. Moreover, it is possible—this is the content of the theorem of Linden et al. (2009)—to show that for certain sorts of initial states of the combined system (namely, those that are superpositions of energy states to which a large number of energy levels contribute significantly, where what counts as "large" is: large compared to the Hilbert-space dimension of the subsystem), the state of the subsystem equilibrates, in the sense that the reduced state of S approaches some quasi-stable state and subsequently spends most of its time near it, with only rare fluctuations away from it.

A result such as this deals with long-term average behaviour of all initial states satisfying the specified condition. That tells us quite a bit, but, by itself, it doesn't tell us *anything at all* about what to expect in the immediate future. Some initial non-equilibrium states will head further away from equilibrium before heading back towards it; others will exhibit the sort of behaviour that experience leads us to expect. To apply equilibration considerations to any actual systems, we need to be reasonably confident that the state we have prepared is not one of those odd states that initially move further away from equilibrium before heading towards equilibrium.

Thus, results such as these, if they are to guide our expectations about near-term behaviour of the systems, require supplementation with considerations about the sorts of pure states that we, or nature, will be able to prepare. We don't want to *completely* exclude states that exhibit bizarre behaviour; what is wanted is that they be deemed improbable. We require, it seems, some sort of epistemic probability distribution over the quantum states of a system at a given time. The prospects for prescribing unique credences over possible initial states are no better in the quantum realm than in the classical. What we can, however, hope for is that differences between reasonable credences will wash out.

Similar considerations apply to studies of the process called *decoherence*. These concern systems that interact with a large, noisy environment, and the tendency for the reduced density operator for such a system to approach a mixture of quasi-classical states. The dynamics of the joint system are unitary and reversible. For that reason, they cannot lead to decoherence for arbitrary initial states of the composite consisting of the system under study and its environment. Models of decoherence typically start with a state in which there are no correlations between system and environment, and evolve it forward. During the process, entanglement builds up between the system and its environment, but it is expected that the correlations that emerge are largely irrelevant to the subsequent evolution of the system *S*. As in the case of equilibration, states in which the process of decoherence is reversed ought not to be regarded as *impossible*, but only unlikely to be realized.

The phenomenon of decoherence raises the same questions about irreversibility raised by equilibration. The unitary evolution involved is invariant under temporal reversal. Thus, it can't be the case that arbitrary initial states of system + environment induce decoherence of the system. What makes the demonstrations of decoherence work is that one starts with a state in which the system and environment are unentangled. Forward evolution entangles the system with its environment, and the originally pure state of the system of interest becomes a mixed one. If one were to run the evolution backwards in time, one would get the same result in the backwards direction. But typically, models of decoherence are meant to be models of situations in which a quantum system has been subject to some preparation procedure which effectively screens off any correlation it might have with its environment.

That preparations of this sort are possible is a widespread assumption of scientific experimentation. It is worth asking what the grounds are for an assumption of this sort. After the preparation, it remains true that the system and the environment share a common past. And, if the dynamics are deterministic and invertible, it will simply not be true that there are *no* variables of system and environment that are correlated with each other. What happens, instead, is that these correlations become buried so deeply in the states of the system that they become essentially irrelevant to future evolution of macrovariables.

The application to actual systems of results about equilibration or decoherence in quantum mechanics, even if they concern states that are initially pure, will proceed much as the sorts of classical results that we have been considering. We begin with a class C of reasonable credences over the initial state of the system, which contain some uncertainty about the initial state, due to limitations of precision of state preparation. The goal should be to track the evolution of these epistemic mixtures, and show that the dynamics takes all mixtures in C to states that yield effectively the same probabilities for outcomes of feasible experiments.

9.3 The measurement problem

9.3.1 The measurement problem formulated

If quantum theory is meant to be (in principle) a universal theory, it should be applicable, in principle, to all physical systems, including systems as large and complicated as our experimental apparatus. Consider, now, a schematized experiment. Suppose we have a quantum system that can be prepared in at least two distinguishable states, $|0\rangle_S$ and $|1\rangle_S$. Let $|R\rangle_A$ be a ready state of the apparatus, that is, a state in which the apparatus is ready to make a measurement.

If the apparatus is working properly, and if the measurement is a minimally disturbing one, the coupling of the system S with the apparatus A should result in an evolution that predictably yields results of the form

$$|0\rangle_S|R\rangle_A \;\Rightarrow\; |0\rangle_S|\text{``0''}\rangle_A$$
$$|1\rangle_S|R\rangle_A \;\Rightarrow\; |1\rangle_S|\text{``1''}\rangle_A$$

(9.1)

where $|\text{``0''}\rangle_A$ and $|\text{``1''}\rangle_A$ are apparatus states indicating results 0 and 1, respectively. Now suppose that the system S is prepared in a superposition of the states $|0\rangle_S$ and $|1\rangle_S$.

$$|\psi(0)\rangle_S = a|0\rangle_S + b|1\rangle_S,$$

(9.2)

where a and b are both non-zero. If the evolution that leads from the pre-experimental state to the post-experimental state is linear Schrödinger evolution, then we will have

$$|\psi(0)\rangle_S|R\rangle_A \;\Rightarrow\; a|0\rangle_S|\text{``0''}\rangle_A + b|1\rangle_S|\text{``1''}\rangle_A. \qquad (9.3)$$

This is not an eigenstate of the instrument reading variable, but is, rather, a state in which system and apparatus are entangled with each other. The eigenstate-eigenvalue link, applied to a state like this, does not yield a definite result for the instrument reading. The problem of what to make of this is called the "measurement problem."

9.3.2 Approaches to the measurement problem

If quantum state evolution proceeds via the Schrödinger equation or some other linear equation, then typical experiments will lead to quantum states that are superpositions of terms corresponding to distinct experimental outcomes. It is sometimes said that this is in conflict with our experience, according to which experimental outcome variables, such as pointer readings, always have definite values. This is a misleading way of putting the issue, as it is not immediately clear how to interpret states of this sort as physical states of a system that includes experimental apparatus, and, if we can't say what it would be like to observe the apparatus to be in such a state, it makes no sense to say that we never observe it to be in a state like that!

Nonetheless, we are faced with an interpretational problem. If we take the quantum state to be a complete description of the system, then the state is, contrary to what we would antecedently expect, not a state corresponding to a unique, definite outcome. This is what led J. S. Bell to remark, "Either the wavefunction, as given by the Schrödinger equation, is not everything, or it is not right" (Bell 1987a: 41; 1987b; 2004: 201). This gives us a (*prima facie*) tidy way of classifying approaches to the measurement problem:

   I. There are approaches that involve a denial that a quantum wave function (or any other way of representing a quantum state), yields a complete description of a physical system.

   II. There are approaches that involve modification of the dynamics to produce a collapse of the wave function in appropriate circumstances.

  III. There are approaches that reject both horns of Bell's dilemma, and hold that quantum states undergo unitary evolution at all times and that a quantum state-description is, in principle, complete.

We include in the first category approaches that deny that a quantum state should be thought of as representing anything in reality at all. These include variants of the Copenhagen interpretation, as well as pragmatic and other anti-realist approaches. Also in the first category are approaches that seek a completion of the quantum state description. These include hidden-variables approaches and modal interpretations. The second category of interpretation motivates a research programme of finding suitable indeterministic modifications of the quantum dynamics. Theories that incorporate such modifications are called *dynamical collapse theories*. Approaches that reject both horns of Bell's dilemma are typified by Everettian, or "many-worlds" interpretations.

As already mentioned, I don't think that the project of constructing an adequate theory on which quantum states do not represent anything in physical reality can succeed. Among those who deny the reality of quantum states, the most prominent is the school that goes by the name of *QBism* (for Quantum Bayesianism), who claim that a quantum state is nothing other than the credence of some agents. My own view is that proponents of this view don't sufficiently appreciate the impact of theorems such as those of Pusey et al. (2012) and Barrett et al. (2014) for a view of that sort. See Myrvold (2020c) for defense of this claim.

9.4 Probabilities in de Broglie–Bohm pilot-wave theories

The de Broglie–Bohm pilot-wave theory, also known as Bohmian Mechanics, is a deterministic theory on which experiments have definite outcomes. This is achieved by taking the rhetoric of "wave-particle duality" seriously; on this theory, the quantum state, represented by a wave function, is physically real, but so are particles, whose motion is guided by the wave-function. The number of physicists actively working on the theory has never been large, but even within this group, there is a divide on how to think of the probabilities in the theory. The divide mirrors a divide on how to think of probabilities in statistical mechanics. One group—which we may call the *Bohmian Mechanics* group—includes Shelly Goldstein, Detlef Dürr, Nino Zanghì, and their collaborators. They take it as a fundamental postulate of any physical theory that there be some measure over the possible ways that the world can be that can be used to judge typicality of properties—a property P is typical just in case the set of states having P is overwhelmingly larger, on the given typicality measure, than the set of states that lack it. The other group, which includes Antony Valentini and his collaborators, seeks to understand the probabilities standardly invoked in the theory as the result of a process of relaxation to equilibrium, thought to have taken place in the early universe.

More details on this, below. First, a brief presentation of the theory.

As the quantum state of an n-body system evolves according to the Schrödinger equation, the probability density over positions,

$$\rho(q, t) = |\Psi(q, t)|^2 \tag{9.4}$$

evolves also. (Here we are using q as a variable for a point in $3n$-dimensional configuration space.) Its evolution satisfies a continuity equation of the same form as the density of a conserved fluid flowing around configuration space,

$$\frac{\partial \rho}{\partial t} + \sum_{k=1}^{3n} \frac{\partial}{\partial q_k} j_k = 0, \tag{9.5}$$

where $\mathbf{j}$ is the probability current density on phase space,

$$j_k(q, t) = -\frac{i\hbar}{2m_k} \left(\psi(q, t) \frac{\partial}{\partial q_k} \psi^*(q, t) - \psi^*(q, t) \frac{\partial}{\partial q_k} \psi(q, t) \right). \tag{9.6}$$

The probability density thus acts like a fluid with a velocity field $\mathbf{v}(q, t)$, given by,

$$v_k(q, t) = j_k(q, t) / \rho(q, t). \tag{9.7}$$

The de Broglie-Bohm pilot wave theory posits that such a system actually does consist of n particles, obeying a non-Newtonian law of motion, which requires that the instantaneous velocities of the particles always satisfy (9.7).

The guidance condition (9.7), together with the Schrödinger equation for the system's wave function, defines a flow on the configuration space of the n-particle system. This permits us to evolve probability distributions over configurations of the particles. Let $\omega(q, t)$ be a density function on configuration space. The de Broglie-Bohm law of motion for the particles, (9.7), has the pleasing feature that, *if* for some time t_0,

$$\omega(q, t_0) = |\Psi(q, t_0)|^2, \tag{9.8}$$

then the same holds for all other times: for all times t,

$$\omega(q, t) = |\Psi(q, t)|^2, \tag{9.9}$$

Thus, evolution of the density $\rho = |\Psi|^2$ via the guidance condition meshes with evolution of the wave function via the Schrödinger equation. This property of the probability density $\rho(q, t)$ is called *equivariance*. It is a cousin of invariance of probability distributions.[1]

This raises the question: what sense can we give to talk of probability in this context? We have a deterministic theory, and, recall, it was a commitment to determinism that led Bernoulli, Laplace, et al. to hold that probability was entirely an epistemic matter. But, if the probabilities involved are to be thought of as credences, reflecting our ignorance of the exact positions of the particles, then we need to ask: Why should these credences have anything to do with the standard, Born-rule probability distribution, represented by $|\Psi(q, t)|^2$?

Insight into this matter is found in an article by Detlef Dürr, Shelly Goldstein, and Nino Zanghì (Dürr et al. 1992), whom will we will refer to as DGZ.

They first explain what it might mean, in the absence of the collapse postulate, to talk about the wave function of a subsystem of the Universe (Recall that the collapse postulate is introduced to deal with the fact that, in its absence, at the end of an experiment, the system is not in an eigenstate of the observable "measured", but, rather, is entangled with the apparatus.). This has to do with the *effective wave function* of a system.

Suppose that we have partitioned the universe into a system of interest, S, and everything else, which we will call the environment, E. Let x be a variable that ranges over the possible configurations of S's particles, and let X be its actual (perhaps unknown) configuration. Let y be a variable that ranges over the possible configurations of the environment E, and let Y be its actual configuration.

---

[1] In the case of classical evolution, for a time-independent Hamiltonian, we had a time-independent flow on phase space, and hence it was possible to have an invariant probability distribution. For the de Broglie–Bohm dynamics, except in the special case of a stationary wave function, the velocity field on configuration space changes with time, which gets in the way of there being an invariant probability distribution. But a probability distribution *can* keep up with the changing flow.

Define the *conditional wave function* of S by

$$\psi_Y(x, t) = \Psi(x, Y, t). \tag{9.10}$$

Suppose that the system S is isolated from the environment (or at least effectively so), a necessary condition for S to be in a pure state more than momentarily. In that case, the Hamiltonian is a sum of a Hamiltonian for S and a Hamiltonian for E.

$$\hat{H} = \hat{H}_S + \hat{H}_E. \tag{9.11}$$

Suppose, also, that the universal wave function Ψ has evolved to a form

$$\Psi(x, y, t) = \psi(x, t)\Phi(y, t) + \Psi^\perp(x, y, t), \tag{9.12}$$

where the sets of values of y for which $\Phi(y, t)$ and $\Psi^\perp(x, y, t)$ are non-zero are macroscopically disjoint, meaning that there is some macroscopic function of y—say, a pointer position—whose value distinguishes the supports of $\Phi(y, t)$ and $\Psi^\perp(x, y, t)$. This is the sort of state that would be the result of a measurement whose result is recorded with the pointer position. Suppose also that Y, the actual value of y, is in the support of Φ.

Then the conditional wave function takes the form

$$\psi_Y(x, t) = \psi(x, t)\Phi(Y, t). \tag{9.13}$$

Moreover, as a consequence of the absence of interaction terms between S and E in the Hamiltonian, the conditional wave function obeys the Schrödinger equation with H_S as its Hamiltonian, and the system has, in effect, its own wave function $\psi(x, t)$. Under such conditions, DGZ call ψ the *effective wave function* of the system S.

DGZ then demonstrate that, if the initial configuration Q is chosen at random according to the standard quantum probability measure, then the conditional distribution of X, conditional on the *actual* value of Y, is correctly given by the conditional wave function $\psi(x)$. Moreover, if one were to perform multiple preparations on a large number of distinct subsystems, and select those with the same effective wave functions $\psi(x)$, and then perform the same experiment on each system, with high probability the frequencies of experimental outcomes would closely match the Born-rule probabilities calculated from ψ.

What DGZ have proven, then, is, in their own words,

> that for *every* initial ψ, this agreement with the predictions of the quantum formalism is obtained for *typical*—i.e., for the overwhelming majority of—choices of initial q. And the sense of typicality here is with respect to the only mathematically natural—because equivariant—candidate at hand, namely, quantum equilibrium.
>
> Thus, on the universal level, the physical significance of quantum equilibrium is as a measure of typicality, and the ultimate justification of the quantum equilibrium hypothesis is, as we shall show, in terms of the statistical behavior arising from a typical initial configuration.
>
> (Dürr et al. 1992: 859)

All of this is a valuable, and, indeed, indispensable contribution towards understanding the role of probabilities in Bohmian mechanics. It still leaves us, however, with the same question as was raised in previous chapters, about the standard equilibrium measures invoked in classical statistical mechanics—what is the status of this typicality measure? In what sense is it a "natural" measure?

Here's what DGZ say about this (p. 868).

> we regard the quantum equilibrium distribution **P**, at least for the time being, solely as a mathematical device, facilitating the extraction of *empirical* statistical regularities from Bohmian mechanics ...and otherwise *devoid of physical significance*. (However, as a *consequence* of our analysis, the reader who so wishes can safely also regard **P** as providing a measure of *subjective* probability for the initial configuration Q. After all, **P** could in fact be *somebody's* subjective probability for Q).

We *can* regard the quantum measure as providing a measure of subjective probability. But *must* we, on pain of unreasonableness?

DGZ say that the standard quantum measure is mathematically natural because equivariant. In a similar vein, one could say (and some do; see Lebowitz 1999a: 520; 1999b: S348; 2001: 53) that the equilibrium measures of standard statistical mechanics are mathematically natural because invariant. This still leaves the question of the significance of equivariance and invariance, in their respective contexts. The answer we have advanced, in the previous chapter, for the significance of invariance is that the standard equilibrium measures are singled out as appropriate *for systems that have undergone a process of equilibration*, where equilibration is understood in terms of convergence of measures—a system that has been evolving undisturbed since a given time t_0 has equilibrated by time t if its dynamics are such that all reasonable credence functions about conditions at t_0 yield, when evolved forward via the system's dynamics, effectively the same probabilities about all macroscopically measurable properties. Any attractor measure of this sort would have to be an invariant measure.

This suggests that we take probabilities in the de Broglie-Bohm theory in the same way, as equilibrium probabilities, the result of a process of "washing-out" of initial conditions. And, indeed, this is precisely what has been suggested by Antony Valentini (1991a; 1991b). He shows that an analog of what Tolman (1938) called the "Generalized H-theorem" (which, as you may recall, we met in §8.5.3) holds in Bohmian mechanics. He calls this the "subquantum H-theorem." As in the statistical mechanical case, the proof proceeds by first defining a coarse-graining operation on probability distributions, then demonstrating that the de Broglie-Bohm dynamics lead to an increase of an appropriately modified entropy.

Since that time, there has been an abundance of theoretical work and computer simulations demonstrating the effectiveness of Bohmian dynamics in bringing about convergence to the equivariant distribution (with exceptions; one can also find systems for which convergence does not occur). See Efthymiopoulos et al. (2017) for an overview.

These results, of course, raise the question of the status of the input measures. You can probably guess what my answer is—we should take them to be reasonable credences, and take the probabilities in de Broglie–Bohm theory to be epistemic chances, in the sense of Chapter 5.

As Dürr, Goldstein, Zanghì, and Valentini all stress, there is a deep analogy between equivariance in Bohmian mechanics and equilibrium in statistical mechanics. One common feature is the value of nonequilibrium as a resource. As Valentini (2002a; 2002b; 2002c) has shown, on the de Broglie–Bohm theory, if one had knowledge of the configuration of a system more precise than the quantum equilibrium probabilities allow, this informational nonequilibrium would permit one to perform a variety of tasks that are provably impossible within standard quantum mechanics.

9.5   Dynamical collapse theories: probabilities as objective chances

The collapse postulate, as formulated by von Neumann and Dirac, is a bit mysterious, as it invokes distinct laws of evolution depending on whether a measurement is occurring. Measurements, however, are just physical interactions that have a certain purpose, and it is not at all clear how it could be that physical systems obey fundamentally different laws depending on whether they are involved in a measurement.

It turns out, however, that it is possible to formulate a unified dynamics for microscopic and macroscopic systems (as Ghirardi et al. 1986 put it), that with high probability, closely mimics the ordinary Schrödinger evolution for an isolated system consisting of a small or not-too-large number of particles, and nevertheless, with high probability, suppresses superpositions of macroscopically distinct states. The best known of these is the GRW theory (Ghirardi et al. 1986), which the authors themselves refer to as *Quantum Mechanics with Spontaneous Localization* (QMSL). This theory, however, does not respect the symmetrization and antisymmetrization requirements on systems of identical particles. There is, however, a collapse theory that does; this is the *Continuous Spontaneous Localization* theory (CSL) (Pearle 1989; Ghirardi et al. 1990).

Among non-Everettian theories, that is, theories in which experiments have unique outcomes, dynamical collapse theories have the advantage over hidden-variables theories that they need not involve any action-at-a-distance. Though probabilities of spatially separated events are not independent of each other, it is probabilities of random experimental outcomes that are correlated, and interventions on a system do not affect the probability of any events at a distance. Nor do such theories require any distinguished relation of distant simultaneity; they are compatible with relativistic causal structure (see Myrvold 2016 for further discussion of this point, and Myrvold 2018a; 2019b for a discussion of ontology for such theories). There, are, in fact, relativistic versions of these theories. There is a relativistic version of the GRW theory, which is restricted to a finite number of non-interacting particles (Dove and Squires 1996; Dove 1996; Tumulka 2006). Also, in different ways Bedingham (2011a,b) and Pearle (2015) have extended the CSL model to relativistic quantum field theories, without the restriction to non-interacting theories.

There is no trouble in interpreting the probabilities in these theories; they are objective chances. The laws of motion of these theories are indeterministic. Given a quantum state at a time t_0, the theory does not prescribe a unique state at a future time t_1. What we obtain from the theory is a probability distribution over future states.

What role will these probabilities play in statistical mechanics? As mentioned above, equilibration results in quantum statistical mechanics are invariably couched in terms of unitary, deterministic evolution. The reversibility argument, therefore, applies in full force. It cannot be proven that we will observe relaxation to equilibrium in the near term for *all* initial states; as in the classical case, if there are initial states that evolve towards equilibrium, there are also states that evolve away from it. We ought not to regard such states as impossible, but merely improbable. As in the classical case, one can hope to prove that, given reasonable credence about states at a given time, one ought to expect equilibration in the near term. Stochastic evolution will place some constraints on credences about states. Even if your credences about the initial state of a system were concentrated on a subset of initial conditions that lead to antithermodynamic behaviour, knowing that a stochastic collapse mechanism is in place will fuzz them out.

Can such lawlike fuzzification completely replace epistemic considerations, rendering them entirely superfluous, as suggested by Albert (1994; 2000: ch. 7)? To answer this question requires investigation—and, as far as I know, this has never been done—whether it can be proven, for a suitable class of Hamiltonians that we believe contains realistic ones, that the sort of stochastic dynamics provided by the GRW or CSL theories yields, for macroscopic systems, an approach to standard equilibrium states for all initial states.

If a result of that sort is obtainable, it will not (contrary to what one might expect) take the form of relaxation towards a stationary equilibrium state. The reason for this is that, on theories of that sort, the standard equilibrium states are *not* stationary states. These theories involve (very slight) violation of conservation of energy, which would result in a gradual warming of an isolated body (too slight to be observed by currently feasible experiments). What one could hope to demonstrate is convergence towards, not a stationary distribution, but one that is equivariant under the stochastic dynamics.

There is a worry about this sort of project, mentioned by Albert (2000: 156–9) as having been raised by Larry Sklar and Philip Pearle. This sort of worry has to do with equilibration, or lack thereof, in systems for which we should expect a good approximation to unitary evolution. One such case would be a gas consisting of around 10^5 molecules. Would it have a tendency to spread out, if originally confined to a small region of the available volume? If so, then this tendency cannot be attributed to the stochasticity of GRW dynamics, as, for a system that small, the theory predicts a close approximation to unitary evolution. Another case has to do with spin echo experiments (recall these from §8.5.1), in which a tendency to approach a state that looks macroscopically random can be observed, and, yet, demonstrably, the evolution has been unitary, or close to it; otherwise, it would not be possible to restore the initial state by the reversing pulse. In both cases, there exist states that don't tend to equilibrate, but these are sensitive to small perturbations, and it would be extraordinarily difficult to reliably prepare them, and we don't expect them to occur in nature. There is, it seems, to be an ineliminable role for epistemic uncertainty about initial conditions to play.

9.6 Everettian interpretations: no probabilities, but a working substitute (perhaps)

Everettian approaches reject both horns of Bell's dilemma. On such an approach, the quantum state obtained by taking the state at some time and applying deterministic, unitary evolution, is taken to be correct, and a quantum state is taken to be capable of representing all of physical reality. When an experiment or some other event occurs that, on the usual way of thinking about things, has a plurality of possible outcomes, whose respective probabilities are calculated from the quantum state by the Born rule, on Everettian interpretations, all eventualities are realized. Provided that the outcomes get recorded in macroscopic variables that are subject to decoherence, any opportunity to demonstrate this plurality of outcomes via an experiment that exhibits interference between these terms will quickly be lost, giving rise to what is, in effect, a branching of worlds. You, the experimenter, will have successors that share your memories and differ in the outcomes they perceive, and none of these is privileged as being the unique future you.

This is a deterministic theory. The evolution at all times is the deterministic, unitary evolution, and the dynamics, together with the state at some time, uniquely determines what will happen after the experiment—there will be branching. Nor is there any room for asking, prior to the experiment, "Which outcome will I perceive?"—the answer is that multiple versions of you will perceive all of the outcomes with nonvanishing Born-rule probability.

In anything like any of the usual senses of "probability" it makes no sense, in the context of a branching theory like this, to talk of a probability that this or that experimental result will be the one that occurs (or the one that is perceived) as there is no such thing as *the* experimental result that will occur. They all will occur, on various branches. It is a presupposition of ordinary probability talk that we assign probabilities to the elements of a set of mutually exclusive alternatives, one, and only one, of which is realized. This presupposition seems not to get a grip in the context of a branching theory.

Is this a problem for Everettian approaches?

One attitude that might be taken is that, beyond epistemic uncertainty about the quantum state, there is no need for any talk of probability. It *is* a deterministic theory, after all!

There are two closely related problems associated with adopting such an attitude. One is how one would ever get to know what the quantum state of a system is in the first place. On the usual account, if one wants to know what quantum state to associate with a given preparation procedure, one does a statistical analysis of experiments involving multiple iterations of the preparation procedure. This proceeds as outlined in §2.4. This requires that one treat the Born rule probabilities as chances, about which one can gain information; in particular, it requires that one have conditional credences, conditional on hypotheses about what the quantum state is, that satisfy the Principal Principle with the Born rule probabilities taken as chances.

Another problem has to do with how one might come to accept Everettian quantum mechanics in the first place. A great deal of the evidence that quantum mechanics is getting something right is statistical in nature. This is evidence that quantum mechanics is getting the *chances* of outcomes of experiments right. If this is jettisoned as nonsensical in an Everettian context, one runs the risk of rendering the theory empirically self-undermining.

The problem of either making sense of probabilities of outcomes of experiments or else finding something else that does the same job, in such a way that the statistical evidence counts in favour of quantum mechanics, even when interpreted in an Everettian vein, is known as the *Everettian evidential problem*.[2]

---

[2] The term "evidential problem" stems, I believe, from Myrvold (2005), though this was not the first occasion on which the problem was raised. See Wallace (2006; 2012: ch. 6) and Greaves and Myrvold (2010) for somewhat different approaches to addressing the problem.

A significant step towards dealing with the problem of probability in an Everettian context was taken by David Deutsch (1999), who introduced the use of decision theory into this context. The logic of Deutsch's argument has been clarified by David Wallace (2003; 2007; see Wallace 2012: ch. 5 for his fullest statement of the argument). The question addressed is: How would an agent who accepted Everettian quantum mechanics and knew the quantum state make decisions? Suppose, for example, that an experiment is about to be done, and goods are to be distributed on the various post-experimental branches in a manner dependent on the experimental outcome. The agent is to indicate preferences between various options of post-branching distributions of goods. (Barry Loewer has dubbed such scenarios *brambles*, as the branching analog of gambles.) For example, given a certain quantum state preparation of a spin-1/2 particle, to be followed by a spin "measurement," you might be asked whether you prefer a scenario in which your successors who see a "+" get a reward and those who see "−" get nothing to a scenario in which the rewards are reversed. The Deutsch– Wallace strategy is to argue that, on the basis of a set of reasonable constraints on the agent's preferences between brambles, an agent's choices between actions will maximize a weighted average of utilities across branches, using the Born-rule weights. That is, such an agent's preferences will match those of someone who thought of the experiments in the usual way, as having unique outcomes with the Born-rule weights as chances.

The argument can be extended to permit an agent who is uncertain about what quantum state to associate with a given preparation to update her credences upon receiving statistical information (see Greaves 2007). The idea is to bring in considerations of accuracy as an epistemic utility, and apply Everettian decision theory to the choice of a credence-updating strategy.

Suppose that one accepts the conclusion of the Deutsch–Wallace argument. This still doesn't solve the Everettian evidential problem, as the argument applies only to an agent who already accepts Everettian quantum mechanics. But it does suggest a strategy for dealing with that problem, outlined by Greaves (2007) and developed more fully by Greaves and Myrvold (2010). One can consider an agent who thinks of experiments as branching events, without commitment to Everettian quantum mechanics or any other theory of how the branching proceeds. The idea is to develop a decision theory for such an agent, by imposing reasonable constraints on her preferences between brambles. Further constraints lead to a representation of the agent's credences as credences about *objective branch weights*. Under the conditions of the representation, hypotheses about the values of these objective branch weights can be tested empirically, in much the same way that hypotheses about chances are, with the result that these branch weights can be estimated, independently of any theory about them. If, then, a theory, such as Everettian quantum mechanics, is formulated, which furnishes a prediction about these weights, one can compare the branch weights predicted by the theory to those estimated empirically and thereby confirm or disconfirm the theory.

Another approach to probability is due to Sebens and Carroll (2018). Their approach focuses on a sort of uncertainty that exists in an Everettian context, even if the agent knows the quantum state and how it evolves. After an experiment has been performed and the result recorded, before becoming aware of the result an agent might have a sort of self-locating uncertainty—though she knows that she has counterparts on branches corresponding to each result, she doesn't yet know which sort of branch she is on. This sort of uncertainty has been called *Vaidman uncertainty*, after the discussion in Vaidman (1998). Sebens and Carroll impose conditions on an agent's post-branching credences that have the effect that her credence in a given result should match the Born-rule weight of branches on which that result obtains.

What drives arguments such as the Deutsch–Wallace argument and the Sebens–Carroll argument is the fact that, in an Everettian context, there is nothing except the quantum state that could be as a basis for formulating a decision rule or a rule for assigning credences to branches. If all one has is the quantum state, and, if one is to respect the condition that zero probability be assigned to subspaces orthogonal to the quantum state, then Born-rule probabilities are pretty much the only option. One may have reservations about the precise form of these arguments, but it is not a simple matter to formulate a coherent alternative to taking Born-rule weights to play the role of probabilities.

Let us assume, for the moment the arguments or a suitable modification of them succeed. What, then, will we say about the role of probabilities, or their surrogates, in the theory? Do they eliminate the need for the sorts of concerns that have been the subject of this book, having to do with epistemic limitations on our knowledge of the physical state?

First of all, decoherence is required to get the theory off the ground, and to obtain a branching structure in the first place. As we have already observed, on reversibility grounds it cannot be the case that this occurs for arbitrary initial states. As with equilibration, we should regard the realization of states that fail to decohere, or in which decoherence appears and then is reversed, as not impossible, but only very improbable.

In addition, since equilibration results in the literature typically involve only deterministic, unitary evolution, and the only role that the Born-rule probabilities play in them is in shaping one's expectations about what will be observed when one performs an experiment on the system, the Everettian account of them will be pretty much the same as everyone else's, and, as argued above, will rely on considerations about limitations of control by us, or, indeed, of any physical process, over what states can be reliably produced.

Thus, it seems, on this approach, as on the other major approaches to the measurement problem, there will be a role for something akin to epistemic chances to play.

9.7   Can classical statistical mechanics stand on its own two feet?

Classical statistical mechanics is applicable when quantum effects can be ignored. A sufficient condition for this is a positive Wigner function that evolves according to the Liouville equation (see §9.9.4). In such a case, there is a built-in restriction on the density function—it cannot be one that violates the uncertainty relations. We have argued that equilibration should be regarded as a process of convergence of probability distributions in an appropriately restricted class. Wigner functions have built-in limitations on how sharply focussed they can be on small regions of phase space.

Might it be the case that quantum uncertainty is all that is needed for classical statistical mechanics? Perhaps. This has, indeed, been suggested by David Wallace (2016a), who regards the status of the standard probabilistic posits in classical statistical mechanics as otherwise a bit mysterious. A case can be made that, in a variety of cases (including, crucially, the paradigm case of a dilute gas), the limitations on phase-space density functions imposed by quantum mechanics suffice to guarantee equilibration.

Quantum mechanics can lend probabilities to classical statistical mechanics. But *need* it do so? Is classical statistical mechanics, regarded as a discipline in its own right, intrinsically conceptually incoherent?

Here's another way to ask the question. Classical statistical mechanics, as developed in the latter half of the nineteenth century, achieved some remarkable successes, which made the domains in which it failed so striking. The domains in which classical statistical mechanics failed included the problem of specific heats (identified by Kelvin at the turn of the century as one of two nineteenth-century clouds over the dynamical theory of heat and light), and the problem of thermal equilibrium of matter with radiation (which, of course, led Planck to the quantum postulate). The areas in which the classical theory broke down were clues that classical physics wasn't quite right. But should nineteenth-century physicists have taken the *successes* of classical statistical mechanics as indications that classical physics is not quite right?

I have argued that the answer is no. All that is needed for classical equilibration results to get off the ground is some uncertainty, expressed as a limitation on the range of density functions over initial conditions that could represent the credences of a reasonable agent. From the standpoint of statistical physics, the grounds for this uncertainty are of no consequence. The case of molecules in a gas interacting with each other, yielding density functions that evolve as if they were probability density functions over the states of a collection of hard spheres bouncing off each other (though in reality the gas is nothing of the sort) goes over smoothly to cases of roulette wheels spinning or billiard balls being shaken, where our uncertainty may be much greater than the absolute minimum required by quantum mechanics. Though the world is, indeed, quantum, there remains a place for considerations of epistemic limitations in statistical physics.

9.8 Conclusion

Much of the philosophical literature on statistical mechanics treats of classical statistical mechanics. This is in sharp contrast to the contemporary *scientific* literature on statistical mechanics. Does it make a difference?

When thinking about the foundations of statistical mechanics, we should always bear in mind that classical physics is not a fundamental theory (and, indeed, that we are not, and never have been, in possession of any theory that is a serious candidate for a complete and fundamental theory). Insofar as our interest in classical statistical mechanics is not merely historical, we should make sure that any conclusions we draw from consideration of classical statistical mechanics carry over to the quantum domain.

Recall that the probabilistic turn in statistical mechanics came about as a result of the reversibility argument. Insofar as quantum statistical mechanics deals with unitary evolution, it is subject to reversibility considerations. The founders of statistical mechanics concluded that the lesson of the reversibility argument is that the sorts of physical states that resist equilibration are ones that cannot be produced with equal facility (as Bernoulli would put it) as those that equilibrate. Much of this book has been devoted to making sense of that. Our proposal has been to consider the effect of dynamical evolution on the sorts of credences an agent like us could have about the physical states produced by the processes we study. In this chapter, we have seen that considerations of this sort are needed in the quantum realm, as well.

9.9 Appendix: a brief introduction to the basics of quantum theory

9.9.1 Quantum states and classical states

In classical mechanics, a maximally specific state-description picks out a point in the system's phase space, which, in turn, yields a definite value for all dynamical variables of the system, which are represented by functions on phase space. A probability distribution that assigns any probability other than one or zero to some physical proposition about the value of a dynamical variable is an incomplete specification of the state of the system. In quantum mechanics, things are different. There are no quantum states that assign definite values to all physical quantities, and probabilities are built into the standard formulation of the theory.

Construction of a quantum theory of some physical system proceeds by first associating the dynamical degrees of freedom with *operators*. These form an algebra; that is, there are well-defined notions of addition and multiplication. Multiplication of these operators is unlike multiplication of numbers, in that order of multiplication can make a difference. That is, it will not always be the case that $\hat{A}\hat{B} = \hat{B}\hat{A}$. When the result of multiplying two operators doesn't depend on their order, the operators are said to *commute*. Under standard assumptions about the algebra of operators, we will be able to represent them as operators on an appropriately constructed Hilbert space (see Ismael 2015 for definition of this, if you're not familiar).

A state can be characterized by an assignment of expectation values to physical quantities ("observables"). These are required to be *linear*: that is, if ρ is the function assigning expectation values to observables, for any observables A, B, and any real numbers α, β,

$$\rho(\alpha A + \beta B) = \alpha\rho(A) + \beta\rho(B). \tag{9.14}$$

A complete set of such expectation values is equivalent to a specification of probabilities for outcomes of all experiments that could be performed on the system.

Two physical quantities are said to be *compatible* if there is a single experiment that yields values for them both; these are associated with operators that commute.

A *pure* state, that is, a maximally specific assignment of expectation values, may be represented in a number of physically equivalent ways, for instance by a family of parallel vectors in the Hilbert space (vectors that are non-zero multiples of each other represent the same state), or a projection operator onto a one-dimensional subspace. In addition to pure states, one can also consider non-pure states, called *mixed*. For example, an experimenter might flip a coin, and subject the system to one or the other of a pair of state-preparations, depending on the outcome of the toss. This procedure yields well-defined probabilities for the outcome of any experiment that can be performed on the systems subjected to this procedure, and so counts as a state-preparation in its own right. One way of representing a state is via a *density operator*. These include both pure states (represented by projection operators) and non-pure states.

For physical quantities that can take on a continuous range of values, such as the position or momentum of a particle, we can represent the state via a function on the space of possible values of the quantity. Thus, for a system consisting of n spinless particles, its state can be represented as a function on its $3n$-dimensional configuration space, or, equivalently, as a function on its $3n$-dimensional momentum space. These are often called "wave functions." For a system of particles with spin, we have to include also a specification of spin states. The spin-state of a finite number of particles that are not all spinless can be represented via a vector in a finite-dimensional Hilbert space. The wave function for such a system will assign a vector in that space to each point in the configuration-space or momentum-space of the system.

If a pure state assigns a definite value to a physical quantity A, a vector that represents the state will be an eigenvector of the operator $\hat{A}$. This gives rise to what has been called the *eigenstate-eigenvalue link*, that is, the interpretative principle that, if a system is assigned a state vector that is an eigenvector of some operator representing a physical quantity, then the corresponding dynamical quantity has the corresponding value, and this can be regarded as a property of the physical system.

The noncontroversial core of quantum theory consists of rules for identifying, for any given system, appropriate operators to represent its dynamical quantities, and an appropriate Hilbert space for these operators to act on. In addition, there are prescriptions for associating quantum states with specified preparation procedures, and for evolving the state of system when it is acted upon by specified external fields or subjected to various manipulations. From the quantum state one can calculate probabilities of outcomes of experiments.

9.9.2 Quantum state evolution

9.9.2.1 The Schrödinger equation
The equation of motion obeyed by a quantum state vector is the Schrödinger equation. It is constructed by first forming the operator $\hat{H}$ corresponding to the total Hamiltonian of the system, which represents the total energy of the system. The rate of change of a state vector is proportional to the result of operating on the vector with the Hamiltonian operator $\hat{H}$.

$$i\hbar \frac{d}{dt} |\psi(t)\rangle = \hat{H}|\psi(t)\rangle. \tag{9.15}$$

There is an operator that takes a state at time t_0 into a state at time t; it is given by

$$\hat{U}(t; t_0) = exp(-i\hat{H}(t - t_0)/\hbar) \tag{9.16}$$

This operator is a linear operator that implements a one-to-one mapping of the Hilbert space to itself that preserves the inner product of any two vectors. Operators with these properties are called *unitary operators*, and, for this reason, evolution according to the Schrödinger equation is called unitary evolution.

For our purposes, the most important features of this equation is that it is deterministic and linear. The state vector at any time, together with the equation, uniquely determine the state vector at any other time. Linearity means that, if two vectors $|\psi_1(t_0)\rangle$ and $|\psi_2(t_0)\rangle$ evolve into vectors $|\psi_1(t)\rangle$ and $|\psi_2(t)\rangle$, respectively, then, if the state at time t_0 is a linear combination of these two, the state at any time t will be the corresponding linear combination of $|\psi_1(t)\rangle$ and $|\psi_2(t)\rangle$.

$$a|\psi_1(t_0)\rangle + b|\psi_2(t_0)\rangle \implies a|\psi_1(t)\rangle + b|\psi_2(t)\rangle. \tag{9.17}$$

9.9.2.2 Time reversal in quantum mechanics

In classical mechanics, the class of dynamically possible motions is invariant under time reversal provided that the Hamiltonian is invariant under an operation that leaves position coordinates unchanged and reverses the sign of their conjugate momenta.

In the quantum realm, there is a result known as *Wigner's theorem*, to the effect that any symmetry operation in quantum mechanics must be implementable by an operator that is either unitary or anti-unitary. A unitary operator is one that preserves the inner product of any two vectors in Hilbert space, and is linear. An anti-unitary operator $\hat{V}$ is a mapping of the Hilbert space that maps the inner product of any two vectors to its complex conjugate, and is antilinear:

$$\hat{V}(\alpha|\psi\rangle + \beta|\phi\rangle) = \alpha^* \hat{V}|\psi\rangle + \beta^* \hat{V}|\phi\rangle. \tag{9.18}$$

See Weinberg (1995: ch. 2, appendix A) for an exposition of Wigner's theorem.

If we have a quantum version of a system whose state is characterized by spatial coordinates $\{q_i\}$ and their conjugate momenta $\{p_i\}$, we demand that the time-reversal operation $|\psi\rangle \Rightarrow |\psi\rangle^T$ be such that the expectation values of the operators $\{\hat{q}_i\}$ corresponding to the spatial coordinates be unchanged, and that the expectation values of the operators $\{\hat{p}_i\}$ change sign. If spin is involved, we demand that it change sign, too. It can be shown that these conditions are implemented by an anti-unitary operator that has the effect of complex-conjugating wave functions.

$$\psi^T(q) = \psi^*(q). \tag{9.19}$$

Under this operation, if the Hamiltonian operator is invariant under temporal inversion, then the dynamical laws are also invariant under that operation.

9.9.2.3 The Quantum recurrence theorem

For a quantum system with a discrete set of energy levels, there is a recurrence theorem, just as there is in classical mechanics (Bocchieri and Loinger 1957). But the quantum version is simpler. For one thing, it applies to *all* initial states, whereas the classical version applies to all but a set of measure zero. Second, though, in the classical case, the recurrence time can vary wildly depending on the initial point (so that, though you know that the system it will return to any neighbourhood of its initial state, you don't know exactly when), the quantum version has a uniform recurrence time.

If a system's Hamiltonian has a discrete spectrum, then, for any $\varepsilon > 0$ and any τ, there exists a time $T > \tau$ such that, for any initial state $|\psi\rangle$,

$$\||\psi(t_0 + T)\rangle - |\psi(t_0)\rangle\| < \varepsilon.$$

9.9.2.4 The collapse postulate

Textbook formulations of quantum mechanics usually include an additional postulate about how to assign a state vector after an experiment. In his influential formulation of quantum theory, von Neumann distinguished between two types of processes: Process 1, which occurs upon performance of an experiment, and Process 2, the unitary evolution that takes place as long as no experiment is made (see von Neumann 1932; 1955: §V.1). In Dirac's formulation, the postulate is

> When we measure a real dynamical variable ξ, the disturbance involved in the act of measurement causes a jump in the state of the dynamical system. From physical continuity, if we make a second measurement of the same dynamical variable ξ immediately after the first, the result of the second measurement must be the same as that of the first. Thus after the first measurement has been made, there is no indeterminacy in the result of the second. Hence, after the first measurement has been made, the system is in an eigenstate of the dynamical variable ξ the eigenvalue it belongs to being equal to the result of the first measurement. This conclusion must still hold if the second measurement is not actually made. In this way we see that a measurement always causes the system to jump into an eigenstate of the dynamical variable that is being measured, the eigenvalue this eigenstate belongs to being equal to the result of the measurement.   (Dirac 1935: 36)

Dirac's "jump" has come to be known as *state vector collapse* or *wave-function collapse*, and the postulation of a jump of this sort is called the *collapse postulate*, or *projection postulate*.

If the quantum state vector is thought to represent only a state of belief or knowledge about a physical system, and not the physical state of the system, then one could regard an abrupt shift in the state vector upon measurement as a shift corresponding to incorporating the result of the measurement into one's belief state. Neither von Neumann nor Dirac, however, seem to have thought of it this way. Note that neither expresses the postulate in terms of "observation"; they speak, instead, of "measurement," treated as a physical process, and there is no suggestion that a conscious observer must become aware of the result of the measurement in order for the collapse postulate to apply. A formulation of a version of the collapse postulate according to which a measurement is not completed until the result is observed is found in London and Bauer (1939). They deny, however, that it represents a mysterious kind of interaction between the observer and the quantum system; for them, the replacement of the pre-observation state vector with a new one is a matter of the observer acquiring new information. These two interpretations of the collapse postulate, as either a real change of the physical state of the system, or as a mere updating of information on the part of an observer, have persisted in the literature.

If state vector collapse is to be regarded as a physical process, this raises the question of what physically distinguishes interventions that are to count as "measurements," capable of inducing an abrupt jump in the state of the system, from other interventions, which induce only continuous, unitary evolution. As John S. Bell (1990) has argued, "measurement" is not an appropriate concept to appear in the formulation of any physical theory that might be taken to be fundamental. If, however, one dispenses with the postulate, this gives rise to the so-called "measurement problem."

9.9.3 Quantum statistical mechanics

Given the structural difference between classical and quantum mechanics, it is remarkable how much of the formalism of classical statistical mechanics goes over into quantum statistical mechanics. Much of the transition consists simply of the replacement of a density function ρ representing a probability distribution on a classical phase space with a density operator $\hat{\rho}$. It might, in fact, be possible to write a statistical mechanical textbook in which almost every formula has a dual meaning, and can be read as classical or quantum, and come out correct on either reading. We have already seen instances of formulas of this sort, in (7.18) and (7.19).

The evolution of a quantum density operator for a system with Hamiltonian operator $\hat{H}$, is given by

$$\frac{d}{dt}\hat{\rho} = -\frac{i}{\hbar}[\hat{H}, \hat{\rho}], \tag{9.20}$$

where $[\hat{H}, \hat{\rho}]$ is the commutator of $\hat{H}$ and $\hat{\rho}$,

$$[\hat{H}, \hat{\rho}] = \hat{H}\hat{\rho} - \hat{\rho}\hat{H}. \tag{9.21}$$

As in classical physics, a density operator that is a function of constants of the motion will be stationary in time. In particular, for a time-independent Hamiltonian, any density operator that is a function only of $\hat{H}$ will be stationary.

Comparison of (9.20) with the Liouville equation is facilitated by writing the latter in terms of the Poisson bracket. The Poisson bracket of any two phase-space functions f, g is defined as

$$\{f, g\} = \sum_{i=1}^{N} \left(\frac{\partial f}{\partial q_i} \frac{\partial g}{\partial p_i} - \frac{\partial f}{\partial p_i} \frac{\partial g}{\partial q_i} \right). \tag{9.22}$$

With this in hand, Liouville's equation (7.3) can be written as

$$\frac{\partial \rho}{\partial t} = -\{\rho, H\}. \tag{9.23}$$

There are quantum versions of the microcanonical and canonical distributions. For the microcanonical distribution, consider a narrow band of energy $[E, E + \delta]$, and consider the set of energy eigenstates with energy in that band. If the system has a discrete set of energy levels, and each energy level has at most a finite degree of degeneracy, this will be a finite set of energy eigenstates, and one can define a mixed state that is an equally weighted mixture of all of these.[3]

The quantum version of the canonical distribution is

$$\hat{\rho} = Z^{-1} e^{-\beta \hat{H}}, \tag{9.24}$$

---

[3] The fact that we are here dealing with a finite set of eigenstates, each of which gets an equal weighting, has encouraged the idea that invocation of a quantum-mechanical microcanonical distribution is a mere application of a Principle of Indifference. The fact that we are dealing with a finite set means that complications that arise in attempting to apply the Principle to infinite sets are bypassed. But the fundamental issue remains: without a choice of a set of states to regard as equiprobable, the Principle offers us no advice. In this case, we must ask: Why energy eigenstates, rather than some other set of states? The fact that these are stationary states makes this a candidate for an equilibrium measure, but, if we invoke considerations of this sort, that is, dynamical considerations, we have left the realm of a pure application of Indifference, and are well on the way to asking the question that is relevant: Is this the sort of state that a process of equilibration will lead to?

where, in order that $\hat{\rho}$ have trace one, we set

$$Z = \text{Tr}[e^{-\beta \hat{H}}]. \tag{9.25}$$

The analog of the Gibbs entropy is the von Neumann entropy,

$$S_N[\hat{\rho}] = -k\text{Tr}[\hat{\rho}\log\hat{\rho}]. \tag{9.26}$$

For an isolated system, it is a constant of the motion.

As in the classical case, we can define a coarse-grained density operator $\hat{\bar{\rho}}$ that is a function only of some set of macrovariables of interest, and a corresponding coarse-grained entropy

$$\bar{S}_N[\hat{\rho}] = S_N[\hat{\bar{\rho}}]. \tag{9.27}$$

9.9.4 From quantum to classical

Classical and quantum statistical mechanics are usually dealt with as separate subjects, and we may well ask what the relation is between the two.

There is, I'm afraid, a picture of the relation that seems to be in the back of may people's minds, when thinking about a gas, treated classically. It is often, I think, implicitly assumed that, even if we take quantum indeterminacy into account, we can think of the molecules as effectively classical, as they are sufficiently massive that they can be treated as if they were more-or-less well-localized objects. Quantum mechanics may be needed to account for their internal structure, but the familiar picture of billiard balls bouncing off of each other will not be too far off.

It turns out that this is completely wrong.[4]

Consider a quantum system (which may have some internal structure, in which case we're talking about the centre of mass wavepacket), of total mass m, which is moving freely, i.e. subject to no external influences. Let $\Delta x(t)$ be the wave-packet spread in x (or any one of the three spatial dimensions) at time t, and let $\Delta p_x(t)$ be the spread in momentum. Take t_0 to be the time at which $\Delta x(t)$ has its smallest value. Then the standard quantum mechanical evolution of the wave packet gives us

---

[4] It was David Wallace who first drew my attention to this. See Wallace (2016a: §8) for a similar calculation.

$$\Delta x(t)^2 = \Delta x(t_0)^2 + \left(\frac{\Delta p_x}{m}\right)^2 (t-t_0)^2. \tag{9.28}$$

By the uncertainty relation for position and momentum, Δp_x (which, since we're dealing with force-free evolution, is a constant of the motion) must satisfy,

$$\Delta p_x \geq \frac{\hbar}{2\,\Delta x(t_0)}, \tag{9.29}$$

and so we have

$$\Delta x(t)^2 \geq \Delta x(t_0)^2 + \left(\frac{\hbar}{2m\Delta x(t_0)}\right)^2 (t-t_0)^2. \tag{9.30}$$

It follows from this that,[5] no matter what the minimum spread $\Delta x(t_0)$ is, for all t,

$$\Delta x(t)^2 \geq \left(\frac{\hbar}{m}\right)|t-t_0|. \tag{9.31}$$

This means that the longest period of time during which a freely moving wavepacket can have a spread less than a certain amount d is

$$\Delta t = \frac{2m\,d^2}{\hbar}. \tag{9.32}$$

Now let's put some numbers in. Most of our atmosphere consists of nitrogen molecules, N_2, which have mass of about 28 atomic mass units, or about 5×10^{-26} kilograms. The mean time between collisions, τ, for nitrogen molecules, at room temperature and atmospheric pressure, is about 6×10^{-10} seconds (see Reif 1968, §12.2 for the calculation). We plug that into equation (9.31), using a value of about 7×10^{-34} m²kg/s for $\hbar$. This gives us the result that, after a time equal to the mean time between collisions, the wavepacket has to spread to at least the size

$$\sqrt{\frac{\hbar\tau}{2m}} \approx 2 \times 10^{-9} \text{ metres.} \tag{9.33}$$

---

[5] Hint:

$$\left(\Delta x(t_0) - \frac{\hbar}{2m\Delta x(t_0)}|t-t_0|\right)^2 \geq 0.$$

Consider the fact that molecular dimensions are on the order of a few multiples of the Bohr radius, which is about 5×10^{-11} metres. The wave-spreading that a molecule undergoes in the intervals between collisions is two orders of magnitude larger than the dimensions of the molecules. In the time required for several collisions, we have nothing at all like well-localized wavepackets.

What gives, then? If gases are actually nothing at all like a collection of classical billiard balls colliding, what's the point of all the calculations made in the framework of the kinetic theory of gases, and why do they yield anything at all like accurate results for the macroscopic properties of gases?

The explanation has to do with the nature of the limiting relation between quantum and classical mechanics. Textbooks often give the impression that, for large, massive systems, quantum mechanics yields an approximation to the classical picture of the world. In fact, it yields nothing of the sort, a fact that is vividly illustrated by Schrödinger cat-style experiments. For any experiment like that, a bare quantum-mechanical treatment (i.e., without collapse, and without extra structure of the sort invoked by hidden-variables theories) yields a state that involves superpositions of macroscopically distinct states of affairs.

What we get, from quantum mechanics, in an appropriate limiting regime, is *probability distributions* on classical phase space that evolve approximately in the same way that classical probability distributions do. That is, the classical limit of a quantum state is a classical probability distribution. The pioneers of statistical mechanics who found themselves studying the behaviour of probability distributions on phase space were, unknowingly, probing a deeper level of physical reality, and constructing a more physically realistic picture, than they would have had they confined themselves to state-descriptions in terms of classical microstates.

And this means that all that effort expended in the name of the kinetic theory of gases, to the extent that it attempted to track, not the evolution of the microstate of the gas, but rather, the evolution of a probability distribution over microstates, was not wasted, and could, indeed, yield results that approximate the quantum results.

One tool that is useful in connection with this is the *Wigner function*, or *Wigner quasi-distribution*.[6] Quantum states, notoriously, do not yield a joint

[6] See Case (2008) for an accessible introduction to the Wigner function, and Hillery et al. (1984) and Lee (1995) for more in-depth overviews of the Wigner function and other phase-space quasi-distributions.

probability distribution over all observables; what they give us instead is, for each set of mutually compatible observables (represented by commuting operators), a probability distribution over those observables. In light of that, it is perhaps surprising that, for any quantum state of an n-body system, it is possible to define a real function $W(x_1, \ldots, x_n; p_1, \ldots, p_n)$ over classical phase space that returns a density function for that state's probability distribution over position, when integrated over all momenta, and a density function for the state's probability distribution over momenta, when integrated over all positions (the Wigner function is just one way to do it, but it's the best-known).

Given a quantum state, which can be represented by a position-space wave-function $\psi(x)$, or by its Fourier transform, the corresponding momentum-space wave-function $\phi(p)$, the Wigner function is defined by,[7]

$$W(\mathbf{x}, \mathbf{p}, t) = \frac{1}{h^3} \int e^{-i\mathbf{p}\cdot\mathbf{y}/\hbar}\, \psi(\mathbf{x} + \mathbf{y}/2, t)\, \psi^*(\mathbf{x} - \mathbf{y}/2, t)\, d^3\mathbf{y} \quad (9.34)$$

$$= \frac{1}{h^3} \int e^{i\mathbf{x}\cdot\mathbf{u}/\hbar}\, \phi(\mathbf{p} + \mathbf{u}/2, t)\, \phi^*(\mathbf{p} - \mathbf{u}/2, t)\, d^3\mathbf{u}. \quad (9.35)$$

It is easy to verify that $W(\mathbf{x}, \mathbf{p}, t)$ is real-valued, and that, as advertised,

$$\int W(\mathbf{x}, \mathbf{p}, t)\, d^3\mathbf{p} = |\psi(\mathbf{x}, t|^2\,; \quad (9.36)$$

$$\int W(\mathbf{x}, \mathbf{p}, t)\, d^3\mathbf{x} = |\phi(\mathbf{p}, t|^2\,. \quad (9.37)$$

It does not, however, serve, in general, as a density function for a probability distribution on phase space because it can take on negative values. In special cases—such as a Gaussian wave-function, for example—it *is* positive everywhere.

We can also define a Wigner function for a mixed state, represented by a density operator $\hat{\rho}$.

These functions were introduced into the literature by E. P. Wigner (1932), in an article appropriately entitled "On the Quantum Correction to Thermodynamic Equilibrium," accompanied by a footnote that reads, "This expression was found by L. Szilard and the present author some years ago for another purpose." We should probably call it the *Wigner–Szilard function*, but, in conformity with prevailing usage and with Stigler's Law of Eponymy, we will stick with "Wigner function."

[7] I'm writing this down for a single particle; the extension to the $6n$-dimensional phase space of an n-particle system is straightforward.

$$W(\mathbf{x}, \mathbf{p}, t) = \frac{1}{h^3} \int e^{-i\mathbf{p}\cdot\mathbf{y}/\hbar} \langle \mathbf{x} + \mathbf{y}/2, t| \hat{\rho} |\mathbf{x} - \mathbf{y}/2, t\rangle \, d^3\mathbf{y} \qquad (9.38)$$

$$= \frac{1}{h^3} \int e^{i\mathbf{x}\cdot\mathbf{u}/\hbar} \langle \mathbf{p} + \mathbf{u}/2, t| \hat{\rho} |\mathbf{p} - \mathbf{u}/2, t\rangle \, d^3\mathbf{u}. \qquad (9.39)$$

As the quantum state evolves in time, so will the corresponding Wigner distribution. It is interesting to compare its evolution to the evolution of a classical density function. Recall that this satisfies the *Liouville equation*, which was eq. (7.3).

$$\frac{\partial \rho}{\partial t} + \sum_{i=1}^{N} \left(\frac{\partial \rho}{\partial q_i} \frac{\partial H}{\partial p_i} - \frac{\partial \rho}{\partial p_i} \frac{\partial H}{\partial q_i} \right) = 0. \qquad (9.40)$$

For a single particle subject to a potential $U(\mathbf{x})$, the Hamiltonian is

$$H(\mathbf{x}, \mathbf{p}) = \frac{p^2}{2m} + U(\mathbf{x}), \qquad (9.41)$$

and so a classical density function $\rho(\mathbf{x}, \mathbf{p}, t)$ on its phase evolves according to

$$\frac{\partial \rho}{\partial t} = - \sum_{i=1}^{3} \left(\frac{p_i}{m} \frac{\partial \rho}{\partial x_i} - \frac{\partial U}{\partial x_i} \frac{\partial \rho}{\partial p_i} \right). \qquad (9.42)$$

Now consider a quantum system, with a Hamiltonian operator $\hat{H}$ having the same form as (9.41). The evolution of the Wigner function W for such a system satisfies

$$\frac{\partial W}{\partial t} = - \sum_{i=1}^{3} \left(\frac{p_i}{m} \frac{\partial W}{\partial x_i} - \frac{\partial U}{\partial x_i} \frac{\partial W}{\partial p_i} \right.$$
$$\left. - \sum_{n=1}^{\infty} \left(-\frac{\hbar}{2} \right)^{2n} \frac{1}{(2n+1)!} \left(\frac{\partial^{2n+1} U}{\partial x_i^{2n+1}} \right) \left(\frac{\partial^{2n+1} W}{\partial p_i^{2n+1}} \right) \right). \qquad (9.43)$$

Notice that the first two terms of the right-hand-side of (9.43) are the same as in the Liouville equation. The remaining terms involve increasing powers of $\hbar$ and increasing derivatives of U and W—odd powers only; the even derivatives play no role. In the special case in which third-order and higher derivatives of $U(\mathbf{x})$ are all zero (which will be true for a free particle or a particle in a harmonic oscillator potential), then these terms all vanish and the evolution equation for the Wigner function is just the Liouville equation.

If the remaining terms of (9.43) are negligible compared to the first two, then the evolution of the Wigner function will approximately satisfy the Liouville equation. This will be the case when the Wigner function and/or the potential is relatively smooth, where "relatively smooth" is to be cashed out as meaning that the higher derivatives are such that all but the classical terms of (9.43) are of negligible size. If, in addition the Wigner function is positive, then we will have recovered from the quantum-theory a classical-like object—a function that behaves like a classical phase-space density function.

Thus, in some special cases, a Wigner function acts like a classical phase-space density function. But not every classical phase-space density function can be obtained this way. For one thing, since the Wigner function has to yield quantum probabilities for position and momentum as marginals, no Wigner function can violate the Heisenberg uncertainty relation,

$$V(X)\,V(P) \;\geq\; \frac{\hbar^2}{4}. \tag{9.44}$$

It can also be shown that there is a bound on the value of Wigner functions. For a Wigner function on the $6n$-dimensional space of n particles,

$$|W(q,p)| \leq \left(\frac{2}{h}\right)^{3n}. \tag{9.45}$$

This is not a restriction that arbitrary density functions are obliged to respect.

Studies of the classical, or quasi-classical, limit of quantum mechanics often emphasize the role of environmentally induced decoherence. Though the quantum state of an isolated system evolves unitarily, and hence a pure state remains pure, a system that interacts with the outside world becomes entangled with it, and its reduced state—that is, the restriction to the system of the global state of the system plus its environment—will become a mixed state. For the right sort of interactions with a large, complex environment, typical initial states lead to states that satisfy some criterion of classicality. One such criterion is positivity of the Wigner function.

10
Epilogue

We began with a puzzle, the puzzle of predictability, which was: Given the vast gulf between what we know about the world, and the amount that would be required to derive predictions on the basis of a specification of the current state, how do we manage to make predictions at all?

The key, as Maxwell already saw in the nineteenth century, lies in *statistical regularities*. The predictable behaviour of macroscopic objects is of the same sort as the predictable coarse-grained regularities that inspired awe in Quetelet and Buckle. The aggregate behaviour of large numbers of molecules or large numbers of people is predictable, not in spite of the exquisite sensitivity of the behaviour of individual humans or molecules, but *because of it*—it is this sort of sensitivity that would make it difficult to create a device that would reliably put the molecules in a box in a state such that it spontaneously divided itself into warmer and cooler halves.

It is a characteristic trait of statistical regularities that exceptions to the expected behaviour are to be regarded, not as impossible or contrary to the fundamental laws of nature, but rather, *highly improbable*. This means that we have to make sense of probability as applied to the physical world. One sort of probability that might be relevant to natural science is objective chance. If the fundamental laws are not deterministic, but chancy, then probabilistic concerns enter in at the most fundamental level.

If this were the only way to make sense of chance-talk applied to the physical world, then, it seems, the successful application of the notion of chance, first to games of chance and then to statistical mechanics, would have counted as evidence against determinism, even prior to the anomalies that led to the realization that classical physics is not quite right and needs to be replaced by a quantum theory. But this is too quick; one should ask whether there is a notion of chance—or something like it—that makes sense in the context of deterministic physics.

Two old standard answers are an approach that seeks to define probability in terms of ratios of numbers of possibility, and an approach that seeks to define probability in terms of frequencies. Objections to these answers

Beyond Chance and Credence. Wayne C. Myrvold, Oxford University Press (2021).
© Wayne C. Myrvold.
DOI: 10.1093/oso/9780198865094.003.0010

are almost as old as the answers themselves, yet both of them have proved surprisingly persistent. For this reason, we have rehearsed the objections to these approaches.

All is not lost, however. There is a concept that does justice to the use of chance-talk in connection with games of chance and statistical mechanics, that makes sense in a deterministic setting. This is a concept that does not fit into the familiar dichotomy of objective chance and epistemic credence, a hybrid concept that combines epistemic considerations regarding limitation of knowledge with considerations of physical dynamics, the concept that I have been calling *epistemic chance*. I have argued that this concept is well suited to resolving some otherwise puzzling features of the use of probabilities in statistical mechanics, and that on the Maxwellian view of thermodynamics, it is exactly the sort of notion that one would expect to employ in any effort to recover appropriately statistical versions of the laws of thermodynamics on the basis of mechanics plus probabilistic considerations.

Just as it goes beyond the familiar dichotomy of chance and credence, epistemic chances go beyond the familiar dichotomy of objective and subjective. A demon with no limitations on epistemic access to the world would have no need of such a concept, and, if the demon were completely unconcerned with the knowings and doings of beings like us, could give a complete account of the world without any mention of epistemic chances. In that sense epistemic chances are tied up with the notion of agency of bounded beings such as ourselves. Yet they are not to be thought of as subjective, either, in the sense of being vulnerable to the idiosyncrasies of individual agents.

I see a kinship between these sorts of considerations, and others that appear in the philosophical literature. A recurrent theme of the work of Jenann Ismael (see, in particular, Ismael 2007; 2016) is to urge a distinction between a "God's-eye view" of the world and the perspectives of agents like us, situated within the world and interacting with it, with limitations on both epistemic access and control of the world. In similar vein, Richard Healey (2017a; 2017b; 2020) has argued that quantum states should be thought of as *informational bridges* whose use is to inform an agent's expectations about physical situations, given certain backing conditions.

Considerations of this sort go beyond the old dichotomy of subjective and objective. Think of Maxwell's account of the work/heat distinction. A complete account of the world in physical terms may employ the concept of energy, but in it will be found no distinction between these two modes of energy transfer. A being with no limitations on its information about the world or its ability to manipulate it would have no need of the distinction.

Neither would a being with no ability to manipulate the world. For beings like us, however, the distinction between energy available to do work and energy dissipated as heat is vitally important. This is not a subjective matter in the sense of being beholden to individual idiosyncracies the way differences in taste are. Their usefulness is predicated on the very mode of being of an agent of bounded capacities situated in the world.

Epistemic chances are of a similar flavour. A complete account of the way the world is need not invoke them. They are, however, significant for agents such as ourselves, situated in a world to which they have some epistemic access that, as Laplace put it, remains infinitely removed from complete knowledge.

11
Appendix: Probability Basics

11.1 Axioms of probability

We assume we have some set S of propositions, which is closed under Boolean operations (and, since $\{\vee, \sim\}$ is a truth-functionally complete set of connectives, it suffices that it be closed under disjunction and negation). A set of propositions that is closed under Boolean operations is called an *algebra* of propositions. A probability assignment is a function assigning a real number $\Pr(p)$ to every proposition $p \in S$, satisfying,

    I. For all p, $\Pr(p) \geq 0$.
    II. If p is logically true, $\Pr(p) = 1$.
    III. If p and q are incompatible, then $\Pr(p \vee q) = \Pr(p) + \Pr(q)$.

A *conditional probability* is a probability of some event, conditional on another event's occurring. The conditional probability of p, conditional on q, is written $\Pr(p \mid q)$, which may be read as "the probability of p, given q." These are related to unconditional probabilities by

$$\Pr(p \,\&\, q) = \Pr(p \mid q)\,\Pr(q). \tag{11.1}$$

If $\Pr(q)$ is greater than zero, then the unconditional probabilities $\Pr(q)$ and $\Pr(p \,\&\, q)$ uniquely determine, via (11.1), the conditional probability $\Pr(p \mid q)$. If, however, $\Pr(q)$ is equal to zero, then the unconditional probabilities $\Pr(q)$ and $\Pr(p \,\&\, q)$ leave it entirely open what, if any, value is to be given to the conditional probability $\Pr(p \mid q)$. Moreover, there may, as has been emphasized by Alan Hájek (2003), be circumstances in which we take the unconditional probabilities $\Pr(q)$ and $\Pr(p \,\&\, q)$ to be vague or ill-defined, and nonetheless may want to take the conditional probability of p given q to be well-defined. Considerations such as these have motivated some authors to take conditional probability as a primitive, rather than defined, notion,

Beyond Chance and Credence. Wayne C. Myrvold, Oxford University Press (2021).
© Wayne C. Myrvold.
DOI: 10.1093/oso/9780198865094.003.0011

and to modify the axioms of probability appropriately. For more on this, see Easwaren (2016) and references therein.

Introducing probabilities conditional on zero-probability conditions introduces complications that can be avoided if they are left undefined; see Myrvold (2015) for a discussion of some of these complications. The usual approach, adopted by the preponderance of textbooks in probability theory, is to leave $\Pr(p|q)$ undefined when q has zero probability. In this book, we will have no need to invoke probabilities conditional on zero-probability conditions. We will, however, adopt as our fourth axiom of probability a version that is neutral with respect to the choice of whether or not to leave such conditional probabilities undefined.

IV. For those p, q for which $\Pr(p \mid q)$ is defined,

$$\Pr(p \,\&\, q) = \Pr(p \mid q)\,\Pr(q).$$

It is easy to show that, from these four axioms follow many other familiar facts about probability functions. Some examples: it follows from the above axioms that, for all p, q,

- If p is logically false, then $\Pr(p) = 0$.
- $\Pr(p) + \Pr(\sim p) = 1$.
- $\Pr(p \,\&\, q) \leq \Pr(q)$.
- $\Pr(p) \leq 1$.
- $\Pr(p \lor q) = \Pr(p) + \Pr(q) - \Pr(p\&q)$.
- If $\Pr(p) = 0$, then $\Pr(p \,\&\, q) = 0$.
- If $\Pr(p) = 1$, then $\Pr(p \,\&\, q) = \Pr(q)$.
- If $p \vDash q$, then $\Pr(p \,\&\, q) = \Pr(p)$.
- If $p \vDash q$, then $\Pr(p) \leq \Pr(q)$.
- If $q \vDash p$ and $Pr(q) > 0$, then $Pr(p \mid q) = 1$.

For many purposes, it is useful to consider a set of propositions that is closed, not only under finite iterations of Boolean operations, but also under infinitary operations. For any countable sequence $\{p_k, k = 1, 2, \ldots\}$ of propositions, let

$$\bigvee_{k=1}^{\infty} p_k$$

be the proposition that is true if and only if at least one of $\{p_k\}$ is true. A *σ-algebra* of propositions is a set of propositions that is closed under negation and disjunction of countable subsets. (It follows from this that it is closed under countable conjunctions, also.)

In probability theory, the set of propositions that are assigned probabilities is usually taken to be a σ-algebra, and probability assignments are taken to satisfy a continuity condition, which we will now explain.

For any sequence $\{p_k\}$ of propositions, consider the finite disjunctions

$$P_n = \bigvee_{k=1}^{n} p_k.$$

For any probability assignment, $\{Pr(P_n)\}$ is a nondecreasing sequence of real numbers, all of which are less than or equal to one. This sequence of numbers, therefore, has a limit, which is the least upper bound of the set of all elements of the sequence. The continuity condition we will impose is that the probability assigned to the infinite disjunction of all of the elements of the set be equal to the limit of the sequence $\{Pr(P_n)\}$.

Axiom of Continuity. For any sequence $\{p_k\}$ of propositions,

$$\mathrm{Pr}\left(\bigvee_{k=1}^{\infty} p_k\right) = \lim_{n \to \infty} \mathrm{Pr}\left(\bigvee_{k=1}^{n} p_k\right).$$

It is easy to show that this is equivalent to the following, which is the form it usually takes, and which we will adopt as our fifth axiom of probability, the *Axiom of Countable Additivity.*

V. If $\{p_k\}$ is a sequence of mutually exclusive propositions, then

$$\mathrm{Pr}\left(\bigvee_{k=1}^{\infty} p_k\right) = \sum_{k=1}^{\infty} \mathrm{Pr}(p_k).$$

11.1.1 Regularity

Notice that the second axiom does *not* say that a proposition is assigned probability 1 *only if* it is logically true, and that the first on our list of consequences of the axioms does not say that $\mathrm{Pr}(p) = 0$ *only if* p is logically false. It is standard practice in probability theory to permit assignments of probability zero to propositions that are *not* logically false.

For example: suppose that a real number is chosen from the interval $[0, 1]$, in such a way that the probability of the chosen number being in any sub-interval $[a, b]$ is equal to the length of that sub-interval. Suppose the algebra of propositions to which we ascribe probabilities contains all propositions of the form "the chosen number is in the interval $[a, b]$," for any numbers a, b, with $a < b$. Closure under complementation and union requires that the algebra also contain propositions specifying the exact value of the chosen number, that is, propositions of the form, "the chosen number is a." As any number is contained in arbitrarily small intervals, the probability assigned to all propositions of that form must be zero.

Some find this counterintuitive, and in some circles it is common to impose a requirement of *regularity*, which is the requirement that every possibility be assigned a non-zero probability. This is achievable if the algebra of propositions is generated by a countable partition of the space of possibilities. If we try to impose it on a probability assigned to an algebra of the sort just considered, we run into a problem. We have to assign a probability to every point value of the chosen number. It is easy to see that at most countably many of these can be assigned a real number greater than zero. To see this, consider: at most one point in the interval can be assigned a probability greater than $1/2$; if one of them has probability greater than $1/2$ of being chosen, all the others must have probability less than $1/2$. Similarly, at most two points can be assigned a probability greater than $1/3$, at most three can be assigned a probability greater than $1/4$, and so on.

So, for any probability assignment, and any n, the set of points assigned a probability greater than $1/n$ is a finite set. Call this set G_n. Now consider the set G, which is the union of all the G_ns. This is the set of all points that are assigned a probability that, is greater than $1/n$ for some n. It is a union of countably many finite sets, and so is a countable set. Its complement is, therefore, an uncountable set, whose members are all points that are assigned probabilities that are less than $1/n$ for all n. If the probabilities to be assigned are real numbers, the only way to do this is to assign them probability zero. The standard response is to reject regularity. Another option is to extend the range of our probability function, and assign some propositions a non-zero *infinitesimal* probability.

Mathematicians, being clever people, have devised a number of different, and non-equivalent, ways to supplement the real numbers by infinitesimals. On each of these, each real number is surrounded by a nimbus of numbers that differ infinitesimally from it. If one wishes to impose regularity via probability functions in some extension of the reals, this should be *done* and not merely gestured at; an advocate of such a view should specify the structure of the extension of the reals that the probability function is to have values in. Too often one has a vague invocation of infinitesimal probabilities with no indication about how they are to be realized.[1]

For our purposes, we will have no need of infinitesimal probabilities. I agree with Good (see §2.2.4) that it is a bit of a joke to pretend that credences are point-valued; to take real-valued credences to be insufficiently precise and require infinitely more precision than a real number affords is to take the joke too far.

11.2 Random variables

A *discrete random variable* is a quantity X that can take on values from a countable set $\{x_i\}$ of real numbers, which may be finite or infinite, with probabilities p_i.

We define the *expectation value* of X by,

$$\text{Exp}(X) = \sum_i p_i x_i, \tag{11.2}$$

provided that the sum converges; if it does not, then the expectation value does not exist (obviously, if the set of potential values is finite, the sum converges). We will also write $\langle X \rangle$ for the expectation value of X.

The *variance*, or *dispersion*, of X is the expectation value of the square of the distance between X and its expectation value.

$$V(X) = \left\langle (X - \langle X \rangle)^2 \right\rangle = \langle X^2 \rangle - \langle X \rangle^2. \tag{11.3}$$

We also define

$$\Delta X = \sqrt{V(X)}. \tag{11.4}$$

$V(X)$ is a measure of how "spread out" the distribution of X is. It takes its minimum value of 0 in the special case in which all the probability is concentrated on a single value in X's range.

---

[1] For a notable exception, see Benci et al. (2018).

Given two random variables X, Y, we define the *covariance* of X and Y by

$$\text{Cov}(X, Y) = \langle (X - \langle X \rangle)(Y - \langle Y \rangle) \rangle$$
$$= \langle XY \rangle - \langle X \rangle \langle Y \rangle. \tag{11.5}$$

We say that X and Y are *positively correlated* if $\text{Cov}(X, Y) > 0$, *negatively correlated* if $\text{Cov}(X, Y) < 0$, *uncorrelated*, or *independent*, if $\text{Cov}(X, Y) = 0$.

It can be shown that

$$\text{Cov}(X, Y)^2 \leq V(X)V(Y). \tag{11.6}$$

Thus, if the variance of a random variable X is zero, then its covariance with any random variable Y is also zero (slogan: no variance, no covariance).

If $V(X)$ and $V(Y)$ are both non-zero, we define their *correlation* as

$$\text{Corr}(X, Y) = \frac{\text{Cov}(X, Y)}{\sqrt{V(X)V(Y)}}. \tag{11.7}$$

This lies within the interval $[-1, 1]$.

11.3 Probability and measure

The modern approach to the theory of probability, which finds its first definitive presentation in the book of Kolmogorov (1933), situates it within the domain of measure theory.

Measure theory generalizes familiar notions such as the size of discrete sets and of areas and volumes of regions of two- or three-dimensional space. One begins with some set Ω, and some set $\mathcal{F}$ of subsets of Ω, which are the ones we are going to ascribe measures to. These could, but need not, include all subsets of Ω. The subsets of Ω that are assigned a measure are the *measurable sets*. They are assumed to form a σ-algebra. To say that a set $\mathcal{F}$ of subsets of Ω is a σ-algebra means:

I. $\Omega \in \mathcal{F}$.
II. $\mathcal{F}$ is closed under complementation. That is, for any $A \in \mathcal{F}$, the complement of A, consisting of all elements of Ω not in A, is also in $\mathcal{F}$.

III. $\mathcal{F}$ is closed under countable unions. That is, for any countable set $\{A_k, k = 1, 2, \ldots\}$ of sets in $\mathcal{F}$, the union of all members of this set is also in $\mathcal{F}$.

Given any set Ω and some set $\mathcal{S}$ of subsets of Ω, there is always a minimal σ-algebra containing all elements of $\mathcal{S}$, which is the unique σ-algebra that consists of precisely those sets that are the members of *every* σ-algebra that contains all members of $\mathcal{S}$. This is called the *σ-algebra generated by $\mathcal{S}$*.

A *measurable space* is a pair $\langle \Omega, \mathcal{F} \rangle$, consisting of a non-empty set Ω, and a σ-algebra $\mathcal{F}$ of subsets of Ω. A *measure* on the measurable space $\langle \Omega, \mathcal{F} \rangle$, is a countably additive set function μ on $\mathcal{F}$ that assigns, to every element of $\mathcal{F}$, either a non-negative real number, or the value ∞ (indicating a set of infinite measure). A measure μ on $\langle \Omega, \mathcal{F} \rangle$ is *finite* if $\mu(\Omega)$ is finite (and hence, the measure assigned to any set in $\mathcal{F}$ is finite); it is a *probability measure* if $\mu(\Omega) = 1$ (from which it follows that all measurable sets are assigned measures in the interval $[0, 1]$).

An important special case of a measurable space is the real line $\mathbb{R}$, and the σ-algebra generated by the set of all intervals (you can take these to be open, half-open, or closed, as these generate the same σ-algebra). This σ-algebra is the *Borel* algebra, $\mathcal{B}$. This generalizes to $\mathbb{R}^n$, which is the set of all ordered n-tuples of real numbers. For any numbers $a_1, \ldots, a_n, b_1, \ldots b_n$, with $a_i < b_i$, there is a "rectangle" consisting of all elements of $\mathbb{R}^n$ with $x_1 \in [a_1, b_1]$ and $x_2 \in [a_2, b_2]$ and $\ldots, x_n \in [a_n, b_n]$. The set of Borel subsets of $\mathbb{R}^n$ is the σ-algebra generated by these rectangles.

Don't be confused by the fact that, in the preceding subsection, probability functions took propositions as arguments, and whereas now they assign numbers to sets. One can take the set X to be a set of mutually exclusive possibilities, one of which is realized; any proposition indicating something about which possibility is realized will pick out a set of such possibilities.

Readers may be wondering why we don't always take the measurable sets to be *all* subsets of the set Ω. For example: for some it may seem intuitively obvious that it makes sense to ascribe a measure to every subset of $\mathbb{R}$. The answer is that, for many interesting cases, there simply is no measure that possesses certain desirable properties and is defined for all subsets. For example, we may wish to have a measure on the real line that ascribes to each interval $[a, b]$ its length $b - a$. Such a measure can be extended to the Borel sets, and the resulting measure is invariant under translations; that is, the measure of any set is unchanged by shifting all its elements by the same amount. Perhaps surprisingly, there is no measure that assigns finite non-zero numbers to bounded intervals, is invariant under translations, is countably additive, and is defined on all subsets of $\mathbb{R}$ (an illustration of *Jagger's Theorem*: You Can't Always Get What You Want). If one is willing to give up countable additivity, it is possible, in one and two dimensions, to define an additive set function that is invariant under translations, as Banach (1923) showed. In three-dimensional space even this is impossible; there is no finitely additive set function that is defined on all subsets of three-dimensional space and is invariant under translations and rotations. This is known as the *Banach–Tarski paradox*.

11.3.1 Borel measure and Lebesgue measure

A measure of particular interest on subsets of $\mathbb{R}^n$ is *Lebesgue measure*. The sets on which it's defined are called the Lebesgue-measurable sets, which include, but are not exhausted by, the Borel-measurable sets. For convenience, we'll consider Borel and Lebesgue measure on the cartesian plane $\mathbb{R}^2$; generalization to $\mathbb{R}^n$ is straightforward. Borel measure is defined in a fairly obvious way; we assign to each rectangle $[a_1, b_1] \times [a_2, b_2]$ its area $(b_1 - a_1)(b_2 - a_2)$. This has a unique countable additive extension to all Borel sets. This measure assigns probability zero to any line or line-segment. On the principle that every subset of a set of measure zero should have measure zero, we would want all subsets of a line in two-dimensional space to have measure zero. But not all subsets of a line are Borel sets. The set of Lebesgue-measurable sets remedies this; we extend the domain of definition of our measure in such a way that it includes all subsets of any set of measure zero.[2] Not every subset of $\mathbb{R}^n$ is a Lebesgue-measurable set, but, provably, any set that you can specify (that is, any set for which you can write down a description that uniquely picks out that set) is Lebesgue-measurable.[3]

---

[2] See any thorough textbook on probability, e.g. Billingsley (2012: §1.3), for a precise definition of the set of Lebesgue-measurable sets.

[3] This may sound fantastic to some. How, you might reasonably ask, can you claim that something is provable about what *I* can specify?

Here's what I mean. There are subsets of the reals, and of $\mathbb{R}^n$, for any *n*, that are not Lebesgue sets. To prove this requires the Axiom of Choice; there exist models of ZF (that is, Zermelo–Fraenkel set theory, without the Axiom of Choice) that contain only subsets of the reals that are Lebesgue measurable (Solovay 1970).

Now the Axiom of Choice is required to prove the existence of something having a certain property *P* when there is such an embarrassment of riches that you can't specify one particular example—that is, anything you can say, if it is satisfied by something having property *P*, is satisfied by infinitely many things with that property. I'm assuming that any description you can give of a set can be formulated in the language of Zermelo–Fraenkel set theory. A model of ZF will have to contain all such specifiable sets. Hence, since there's a model of ZF that contains only Lebesgue-measurable sets, all the specifiable sets are Lebesgue-measurable.

A word of caution. There is a tendency, in some of the philosophical literature, to talk of "Lebesgue measure" on a classical phase space or some other state space of a physical system. Please refrain from doing so! Given an n-dimensional state space Γ, if there's a coordinate map that covers the space—that is, a smooth, invertible map from Γ into $\mathbb{R}^n$—we can take Lebesgue measure on $\mathbb{R}^n$ and use it to induce a measure on the subsets of Γ that are mapped into Lebesgue-measurable subsets of $\mathbb{R}^n$. If there's one such coordinate map, there are many, as any smooth one-one mapping of $\mathbb{R}^n$ to itself will give us another set of coordinates. Therefore, there can't be a unique measure on Γ that's induced, via coordinatization, by the Lebesgue measure on $\mathbb{R}^n$.

11.3.2 Integrals and density functions

Given a measurable space $\langle \Omega, \mathcal{F} \rangle$, a function $f : \Omega \rightarrow \mathbb{R}$ is said to be measurable if, for every Borel subset of $\mathbb{R}$, its inverse image,

$$f^{-1}(B) = \{ \omega \in \Omega \,|\, f(\omega) \in B \}$$

is in $\mathcal{F}$. Given a measure μ on $\langle \Omega, \mathcal{F} \rangle$, we can define (Lebesgue) integrals

$$\int_\Omega f\, d\mu$$

in the usual way; see a good textbook of calculus if you're not familiar.

If f is a non-negative measurable function, then ρ, defined by

$$\rho(A) = \int_A f\, d\mu \tag{11.8}$$

is a measure on $\langle \Omega, \mathcal{F} \rangle$. We say, in such a case, that f is a density function for ρ with respect to μ.

As long as there are non-empty subsets of Ω that are assigned zero measure by μ, then, if there's one density function for a measure ρ with respect to μ, there are infinitely many, as any two functions that differ only on a set of μ-measure zero will yield the same results. That's why we say f is *a* density function for ρ, rather than *the* density function.

If ρ has a density with respect to μ, then any set of zero μ-measure will also have zero ρ-measure. Another way of saying that is to say that ρ is *absolutely continuous* with respect to μ.

It turns out that, if the measures μ and ρ are finite or merely σ-finite,[4] then the converse holds. If μ and ρ are σ-finite measures, with ρ absolutely continuous with respect to μ, then ρ has a density with respect to μ. This is known as the *Radon–Nikodym theorem*. Its proof can be found in many textbooks of probability theory; see e.g. Billingsley (2012: §32).

Obviously, the same measure ρ may have density functions with respect to two different measures μ and v, and these density functions will, in general, differ. To illustrate this, suppose we have a line, and two coordinatizations of it; which we will call x and u, related by

$$u = U(x). \tag{11.9}$$

where U is a smooth one-one function. As this is a one-one function, it has an inverse,

$$x = X(u). \tag{11.10}$$

Let λ_x and λ_u be the Lebesgue measures on the line induced by x and u, respectively. Suppose that we have a measure ρ that has a density function $f(x)$ with respect to λ_x, and a density function $g(u)$ with respect to λ_u. These two functions are related by the conditions that

$$g(u) = f(X(u)) X'(u);$$
$$\tag{11.11}$$
$$f(x) = g(U(x)) U'(x).$$

hold for almost all x and u (that is, the exceptions, if any, are a set of zero measure). For instance: the measure λ_x, Lebesgue measure with respect to x, has a density function, with respect to itself, which is equal to 1 everywhere. With respect to λ_u, it has density

$$g(u) = X'(u). \tag{11.12}$$

---

[4] To say that a measure μ on $\langle \Omega, \mathcal{F} \rangle$ is σ-finite means that there is a countable partition $\{A_i\}$ of Ω with each $\mu(A_i)$ finite. Lebesgue measure on $\mathbb{R}^n$, for example, though not finite, is σ-finite.

11.3.3 Random variables, again

Earlier we considered random variables that take on a discrete set of values. This generalizes to random variables that can take on any real number (or any real number in a certain range). Suppose we have a probability space $\langle \Omega, \mathcal{F}, P \rangle$. For any measurable function $f : \Omega \to \mathbb{R}$ there is a random variable F, whose value is $f(\omega)$, for any $\omega \in \Omega$. The expectation value of F is,

$$\langle F \rangle_P = \int_P f \, dP, \tag{11.13}$$

provided that the integral exists; if the integral fails to converge, the expectation value does not exist. The variance of F is, as with discrete random variables, the expectation value of $(F - \langle F \rangle_P)^2$.

11.4 Laws of Large Numbers

There is a family of theorems, dating back to Bernoulli (1713), known as Laws of Large Numbers. We illustrate the basic idea with reference to coin tosses.

Imagine a sequence of tosses of a coin, not necessarily a fair one, with probability p for *Heads* on each individual toss, with the outcomes of distinct tosses probabilistically independent of each other. Let F_n be the relative frequency of *Heads* in the first n tosses, that is, the number of cases in which the coin lands *Heads* in the first n tosses, divided by n. The relative frequency F_n will tend to approximate the single-case probability p for large n, in the sense that, for any interval containing p, no matter how small, we can make the probability that F_n is in that interval as close to one as we like, by taking n sufficiently large. This is one form of the *Weak Law of Large Numbers*. The *Strong Law of Large Numbers* considers, not properties of finite sequences of tosses, but properties of an entire infinite sequence of tosses. It says that, with probability one, the sequence F_n converges to p.

In thinking of these laws, it is helpful to think of the example of *Bernoulli Trials*. Consider a sequence $\{A_i\}$ of propositions, such that each A_i has the same probability p, and they are probabilistically independent of each other. The sequence is a sequence of trials, each of which is referred to a *Bernoulli trial*. We may represent this as a sequence $\{X_i\}$ of random variables, each of which takes on only two values (which we may choose to be 0 and 1), which all have the same distribution and are independent of each other (a special case of *independent, identically distributed* random variables, often abbreviated "i.i.d."). Such a sequence is known as a *Bernoulli process.*

An infinitely long sequence of coin tosses with constant chance of *Heads* is an example of a sequence of Bernoulli trials. For a case like this, we can define a sequence of random variables $\{X_i\}$ by taking X_i to be equal to 1 if the outcome of the ith toss is *Heads*, and equal to 0 if it is *Tails*.

11.4.1 Weak Laws of Large Numbers

Consider a sequence of Bernoulli trials, and let F_n be the relative frequency of one of the outcomes in the first n trials. For any interval containing p but not swallowing the whole unit interval $[0, 1]$ (this could be, for example, the interval $[p - \varepsilon, p + \varepsilon]$ for small ε), the probability that F_n is in the interval increases with increasing n, and can be made as close to unity as one would like by taking n sufficiently large. Moreover, we can give explicit estimates on how large is sufficiently large. The conceptually easiest way to do this is via an inequality that was first enunciated by Irenée Jules Bienaymé and which, in accordance with Stigler's Law of Eponymy, is usually referred to as the *Chebyshev Inequality.*

Lemma 11.1 *Let X be any random variable, with finite expectation value μ and variance σ^2. Then, for any $\varepsilon > 0$,*

$$Pr(|X - \mu| \le \varepsilon) \ge 1 - \frac{\sigma^2}{\varepsilon^2}.$$

Applying this to our coin toss example, let X_i be a random variable that takes on the value 1 if the ith toss comes out *Heads*, and 0 otherwise. Then we have

$$F_n = \frac{1}{n} \sum_{i=1}^{n} X_i. \tag{11.14}$$

Each X_i has expectation value p, and variance $V(X_i) = p(1 - p)$. Moreover, for distinct i, j, X_i is independent of X_j. A simple calculation gives us

$$V(F_n) = \frac{p(1 - p)}{n}. \tag{11.15}$$

Applying the Chebyshev inequality gives

$$Pr(|F_n - p| \leq \varepsilon) \geq 1 - \frac{p(1-p)}{n\,\varepsilon^2} \geq 1 - \frac{1}{4n\,\varepsilon^2}. \qquad (11.16)$$

Obviously, for any given ε, no matter how small, the right-hand side of this inequality can be made as close to unity as one would like by taking n sufficiently large. For example, if we wanted to guarantee probability of at least 0.99 that the relative frequency of *Heads* is within one-tenth of the probability of *Heads* on an individual toss, (11.16) tells us that it suffices to take n greater than or equal to 2,500 (note that we don't need to know the value of p to have this assurance!). This does, indeed, guarantee it, but it's not a very good bound; as a matter of fact, it suffices to take n greater than or equal to 105.

The Chebyshev inequality gives us our first version of the Weak Law of Large Numbers.

Proposition 11.1 *Let $\{X_i\}$ be a sequence of independent random variables, with common expectation value μ, whose variances are all less than or equal to some maximum value K. Let*

$$\bar{X}_n = \frac{1}{n} \sum_{i=1}^{n} X_i.$$

Then, for any $\varepsilon > 0$,

$$Pr(|\bar{X}_n - \mu| \leq \varepsilon) \geq 1 - \frac{K}{n\,\varepsilon^2}.$$

Note that there is no assumption that the random variables $\{X_i\}$ are identically distributed, though this is of course an important special case.

An immediate corollary of Proposition 11.1 is the following.

Proposition 11.2 *Let $\{X_i\}$ be a sequence of random variables satisfying the conditions of Proposition 11.1. Then, for any $\varepsilon, \delta > 0$, there exists N such that*

$$Pr(|\bar{X}_n - \mu| \leq \varepsilon) \geq 1 - \delta$$

for all $n \geq N$.

A sequence of random variables $\{Z_n\}$ is said to *converge in probability* to a limiting random variable Z iff and only if, for every $\varepsilon > 0$,

$$\Pr(|Z_n - Z| \leq \varepsilon) \rightarrow 1 \text{ as } n \rightarrow \infty.$$

We write,

$$Z_n \rightarrow_p Z.$$

Thus, for the conclusion of Proposition 11.2, we can also write,

$$\bar{X}_n \rightarrow_p \mu.$$

Proposition 11.2 is what is often called the *Weak Law of Large Numbers*. Note that, though Proposition 11.2 only states convergence, without giving any clue as to rate of convergence, this is readily supplied by Proposition 11.1, and there are other theorems that give better bounds, subject to various conditions.

The Weak Law of Large Numbers generalizes in various ways. For example, suppose the expectation values of the random variables $\{X_i\}$ are not all equal, but converge to some limit μ. Then we will still have convergence in probability of $\bar{X}_n$ to μ. If the sequence of expectation values doesn't converge to a limit, we can consider the running average

$$\bar{\mu}_n = \frac{1}{n} \sum_{i=1}^{n} \mu_i.$$

What we obtain then is that $\bar{X}_n$ will converge in probability to the running average $\bar{\mu}_n$. The variances need not be uniformly bounded, as long as they don't increase too quickly. As one can see from the Chebyshev inequality, as long as the variances $V(\bar{X}_n)$ go to zero as n goes to infinity, $\bar{X}_n$ will converge in probability to $\bar{\mu}_n$. This means that the variables $\{X_i\}$ need not be strictly independent; we can tolerate some correlation between the variables, as long as it is not too much.

Proposition 11.3 *Let $\{X_i\}$ be a sequence of random variables, with finite expectation values μ_i, and finite variances σ_i^2. Let $c_{ij} = Cov(X_i, X_j)$. If*

$$\frac{1}{n^2} \sum_{i=1}^{n} \sigma_i^2 \rightarrow 0 \text{ as } n \rightarrow \infty$$

and

$$\frac{1}{n^2} \sum_{i=1}^{n} \sum_{j=i+1}^{n} c_{ij} \to 0 \ as \ n \to \infty$$

then, for all $\varepsilon > 0$,

$$Pr(|\bar{X}_n - \bar{\mu}_n| \le \varepsilon) \to 1 \ as \ n \to \infty.$$

This follows straightforwardly from the Chebyshev inequality, from which one can obtain a bound on the rate of convergence in terms of the quantities mentioned.

Although the path we have taken to the Weak Law of Large Numbers, via the Chebyshev inequality, requires the summed random variables to have finite variance, this is not a necessary condition for convergence in probability; as proven by Khintchine (1929), all that is required is that $|X_k|$ have finite expectation value.

Proposition 11.4 *Let $\{X_i\}$ be a sequence of independent, identically distributed random variables, such that $\langle |X_i| \rangle$ is finite. Let $\mu = \langle |X_i| \rangle$, and let*

$$\bar{X}_n = \frac{1}{n} \sum_{i=1}^{n} X_i.$$

Then

$$\bar{X}_n \to_p \mu \ as \ n \to \infty.$$

For proof, see Feller (1957: §X.2).

11.4.2 The Central Limit Theorem

One route to the Weak Law of Large Numbers is via the *Central Limit Theorem*, which has to do with the approximation of a sum of a large number of i.i.d random variables via a normal distribution.

A *normal distribution*, with mean μ and variance σ^2, is one with density function, with respect to Lebesgue measure,

$$f(x) = \frac{1}{\sigma\sqrt{2\pi}} e^{-(x-\mu)^2/2\sigma^2}. \tag{11.17}$$

Let $\{X_k\}$ be a sequence of independent, identically distributed random variables, with expectation value μ and finite variance σ^2. Let S_n be the sum of the first n of them.

$$S_n = \sum_{k=1}^{n} X_k. \tag{11.18}$$

The random variable S_n has expectation value $n\mu$, and variance $n\sigma^2$. $S_n/\sqrt{n}$, therefore, has expectation value $\sqrt{n}\mu$, and variance σ^2. The Central Limit Theorem says that the distributions of $S_n/\sqrt{n}$ approach a normal distributions as n is increased indefinitely. That is, for any real number a,

$$\lim_{n\to\infty} \Pr[(S_n - n\mu)/\sqrt{n} \le a] = \frac{1}{\sigma\sqrt{2\pi}} \int_{-\infty}^{a} e^{-x^2/2\sigma^2} dx. \tag{11.19}$$

The proof can be found in a number of textbooks of probability (e.g. Billingsley 2012: §27; Rosenthal 2006: §11.2). Standard proofs prove convergence without being particularly informative about bounds on the differences between the distributions of $S_n/\sqrt{n}$ and the normal distribution. Such bounds can be obtained under further assumptions about the probability distribution. For example, on the assumption that $|X_k - \mu|^3$ has a finite expectation value ρ (an assumption not needed for the general Central Limit theorem), the Berry–Esseen theorem says that

$$\left| \Pr[(S_n - n\mu)/\sqrt{n} \le a] - \frac{1}{\sigma\sqrt{2\pi}} \int_{-\infty}^{a} e^{-x^2/2\sigma^2} dx \right| \le \frac{C\rho}{\sigma^3\sqrt{n}}, \tag{11.20}$$

where C is a constant originally estimated by Esseen (1956) to be less than or equal to 7.59 (see Feller 1970: §XVI.5). This estimate has since been improved; it can be shown that (11.20) holds with $C = 0.469$ (Shevtsova 2013).

For a sequence of coin tosses, the relative frequency of *Heads* in the first n tosses, F_n, is

$$F_n = \frac{1}{n} \sum_{k=1}^{n} X_k = \frac{1}{n} S_n. \tag{11.21}$$

If, for some n, the distribution for $S_n/\sqrt{n}$ is well approximated by a normal distribution with mean $\sqrt{n}\mu$, and variance σ^2, then the distribution of F_n is well approximated by a normal distribution with mean μ and variance σ^2/n.

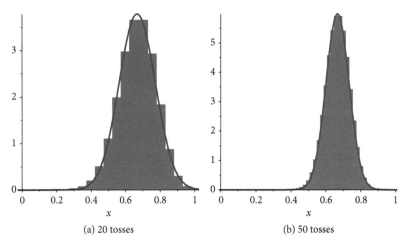

Figure 11.1. Probabilities of possible values for F_n, for $p = 2/3$, with normal approximations

This is illustrated, for a sequence of coin tosses biassed two to one in favour of *Heads*, in Figure 11.1. F_n can only take on a discrete set of values m/n, $m = 0, \ldots, n$. In Figure 11.1, I have plotted a bar chart indicating the probabilities for these values, so that the area of the rectangle whose base is centred on m/n is the probability of getting m heads in n trials, for sequences of 20 and 50 tosses.

As with the Weak Law of Large Numbers, the Central Limit Theorem generalizes to sequences of independent random variables that are not identically distributed. A necessary and sufficient condition for the Central Limit Theorem to hold is the *Lindeberg condition*. We will not state it here; see Feller (1957: §X.5).

11.4.3 The Strong Law of Large Numbers

As mentioned, the *Strong Law of Large Numbers* has to do with properties of an infinite sequence $\{X_k\}$ of random variables (think of an infinitely long string of coin flips). The Weak Law says that, for any interval Δ containing μ, the shared expectation value of all X_k, we can make the probability that $\bar{X}_n$ is in Δ as large as we like by taking n sufficiently large. We might also ask what the probability is that, with increasing n, $\bar{X}_n$ will eventually land in Δ and stay within it forevermore. The Strong Law of Large Numbers says that the probability that it will do this is one. That is, if you were able to bet on propositions concerning properties of an entire infinite sequence of coin flips, you should bet, at arbitrarily long odds, that the sequence $\{\bar{X}_n\}$ will converge to μ.

An immediate consequence of the Strong Law is the *impossibility of an effective gambling system*. Suppose that a coin is to be flipped infinitely many times. You get to choose which flips you are going to bet on. Suppose you adopt a strategy of betting only on certain flips, with the proviso that your decision whether or not to bet on a given flip depends only on the outcomes of flips prior to that one. You might, like so many a naïve gambler, think that, because the relative frequency of *Heads* is almost sure to converge to p, the single-case probability of *Heads*, it follows that, after a long string of *Tails*, an occurrence of *Heads* is more likely, as compensation for the long string of *Tails*. Whatever strategy you adopt for selection, the Strong Law applies to the sequence picked out by your strategy, and, with probability one, the frequency of *Heads* in your subsequence will converge to p. If you have a friend who adopts a different strategy, then, with probability one, the frequency of *Heads* will converge to p in both your subsequence and your friend's. The same holds if you have a countable infinity of friends: with probability one, all the frequencies in all of the sequences picked out by a countable infinity of selection rules will converge to p.

In actual physical applications, we never have to deal with infinite sequences of events, and so the Strong Law, unlike the Weak Law, is of limited interest. It is, however, relevant to the version of frequentism defended by von Mises.

References

Abrams, M. (2012). Mechanistic probability. *Synthese 187*, 343–375.

Ainsworth, P. (2005). The spin-echo experiment and statistical mechanics. *Foundations of Physics Letters 18*, 621–635.

Albert, D. (2000). *Time and Chance*. Cambridge, MA: Harvard University Press.

Albert, D. Z. (1994). The foundations of quantum mechanics and the approach to thermodynamic equilibrium. *The British Journal for the Philosophy of Science 45*, 669–677.

Albert, D. Z. (2015). Physics and chance. In *After Physics*, 1–30. Cambridge, MA: Harvard University Press.

Albrecht, A. and L. Sorbo (2004). Can the universe afford inflation? *Physical Review D 70*, 063528.

Banach, S. (1923). Sur le problème de la mesure. *Fundamenta Mathematicae 4*, 7–33.

Barrett, J., E. G. Cavalcanti, R. Lal, and O. J. E. Maroney (2014). No ψ-epistemic model can fully explain the indistinguishability of quantum states. *Physical Review Letters 112*, 250403.

Bartolotta, A., S. M. Carroll, S. Leichenauer, and J. Pollack (2016). Bayesian second law of thermodynamics. *Physical Review E 94*, 022102.

Bedingham, D. (2011a). Relativistic state reduction model. *Journal of Physics: Conference series 306*, 012034.

Bedingham, D. (2011b). Relativistic state reduction dynamics. *Foundations of Physics 41*, 686–704.

Beisbart, C. and S. Hartmann (eds) (2011). *Probabilities in Physics*. Oxford: Oxford University Press.

Bell, J. S. (1987a). Are there quantum jumps? In C. Kilmister (ed.), *Schrödinger: Centenary Celebration of a Polymath*, 41–52. Cambridge: Cambridge University Press. Reprinted in Bell (1987b, 201–212).

Bell, J. S. (1987b). *Speakable and Unspeakable in Quantum Mechanics*. Cambridge: Cambridge University Press.

Bell, J. S. (1990). Against 'measurement' *Physics World 3*, 33–40. Reprinted in Bell (2004, 213–231).

Bell, J. S. (2004). *Speakable and Unspeakable in Quantum Mechanics* (2nd edn). Cambridge: Cambridge University Press.

Bell, M. and S. Gao (eds) (2016). *Quantum Nonlocality and Reality: 50 Years of Bell's Theorem*. Cambridge: Cambridge University Press.

Benci, V., L. Horsten, and S. Wenmackers (2018). Infinitesimal probabilities. *The British Journal for the Philosophy of Science 69*, 509–552.

Berkovitz, J., R. Frigg, and F. Kronz (2006). The ergodic hierarchy, randomness and Hamiltonian chaos. *Studies in History and Philosophy of Modern Physics 37*, 661–691.

Bernoulli, J. (1713). *Ars Conjectandi, Opus Posthumum*. Basel: Impensis Thurnisiorum, Fratrum.

Bernoulli, J. (2006). *The Art of Conjecturing*. Baltimore, MD: Johns Hopkins University Press. Translated by E. D. Sylla.

Bertrand, J. (1870). *Traité de Calcul Différentiel et de Calcul Intégrale*. Paris: Gauthier-Villars.

Bertrand, J. (1889). *Calcul des Probabilités*. Paris: Gauthier-Villars.

Billingsley, P. (2012). *Probability and Measure, Anniversary Edition*. Hoboken, NJ: Wiley.

Blackwood, E. (1877). Solution of question 5461. *Mathematical Questions, with their Solutions, from the "Educational Times" 28*, 109.

Blackwood, E. (1878). Miss Blackwood's reply to Helen Thomson's verses on "random chords." *Mathematical Questions, with their Solutions, from the "Educational Times" 29*, 62–63.

Blatt, J. M. (1959). An alternative approach to the ergodic problem. *Progress of Theoretical Physics 22*, 745–756.

Bocchieri, P. and A. Loinger (1957). Quantum recurrence theorem. *Physical Review 107*, 337–338.

Boltzmann, L. (1871a). Einige allgemeine Sätze über Wärmegleichgewichte. *Sitzungberichte der Kaiserlichen Akademie der Wissenschaften. Mathematisch-Naturwissenschaftliche Classe 63*, 679–711. Reprinted in Boltzmann (1909a, 259–287).

Boltzmann, L. (1871b). Analytischer Beweis des zweiten Haupsatzes der mechanischen Wärmetheorie aus den Sätzen über das Gleichgewicht der lebengdigen Kraft. *Sitzungsberichte der Akademie der Wissenschaften. Mathematisch-Naturwissenschaften Klasse 63*, 712–732. Reprinted in Boltzmann (1909a, 288–308).

Boltzmann, L. (1872). Weitere Studien über das Wärmegleichgewicht unter Gasmolekülen. *Sitzungberichte der Kaiserlichen Akademie der Wissenschaften. Mathematisch-Naturwissenschaftliche Classe 66*, 275–370. English translation, Boltzmann (1966a).

Boltzmann, L. (1877a). Bemerkungen über einige Probleme der mechanische Wärmetheorie. *Sitzungsberichte der Kaiserlichen Akademie der Wissenschaften. Mathematisch-Naturwissenschaftliche Classe 75*, 62–100. Reprinted in Boltzmann (1909b, 113–148). English translation of §2 in Boltzmann (1966b).

Boltzmann, L. (1877b). Über die Beziehung zwischen dem zweiten Hauptsatze der mechanischen Wärmetheorie und der Wahrscheinlichkeitsrechnung resp. dem Sätzen über das Wärmegleichgewicht. *Sitzungsberichte der Kaiserlichen Akademie der Wissenschaften. Mathematisch-Naturwissenschaftliche Classe 76*, 373–435. Reprinted in Boltzmann (1909b, 164–223).

Boltzmann, L. (1895). On certain questions of the theory of gases. *Nature 51*, 413–415.

Boltzmann, L. (1896). *Vorlesungen Über Gastheorie. I. Thiel.* Berlin: Johann Ambrosius Barth.

Boltzmann, L. (1898). *Vorlesungen Über Gastheorie. II. Thiel.* Berlin: Johann Ambrosius Barth.

Boltzmann, L. (1909a). *Wissenschaftliche Abhandlungen. I. Band.* Leipzig: J. A. Barth.

Boltzmann, L. (1909b). *Wissenschaftliche Abhandlungen. II. Band.* Leipzig: J. A. Barth.

Boltzmann, L. (1964). *Lectures on Gas Theory.* Berkeley: University of California Press. Translation of Boltzmann (1896; 1898).

Boltzmann, L. (1966a). Further studies on the thermal equilibrium of gas molecules. In Brush (1966, 88–175). English translation of Boltzmann (1872).

Boltzmann, L. (1966b). On the relation of a general mechanical theorem to the second law of thermodynamics. In Brush (1966, 188–193). English translation of §2 of Boltzmann (1877a).

Bradley, S. (2019). Imprecise probabilities. In E. N. Zalta (ed.), *The Stanford Encyclopedia of Philosophy* (Spring 2019). Stanford, CA: Metaphysics Research Lab, Stanford University.

Brier, G. W. (1950). Verification of forecasts expressed in terms of probability. *Monthly Weather Review 78*, 1–3.

Brown, H. R., W. Myrvold, and J. Uffink (2009). Boltzmann's H-theorem, its discontents, and the birth of statistical mechanics. *Studies in History and Philosophy of Modern Physics 40*, 174–191.

Brown, H. R. and J. Uffink (2001). The origins of time-asymmetry in thermodynamics: The minus first law. *Studies in History and Philosophy of Modern Physics 32*, 525–538.

Brush, S. G. (ed.) (1966). *Kinetic Theory, vol. 2: Irreversible Processes.* Oxford: Pergamon Press.

Brush, S. G. (1976). *The Kind of Motion We Call Heat,* bk 1. Amsterdam: North-Holland.

Butterfield, J. (1996). Whither the minds? *The British Journal for the Philosophy of Science 47*, 200–221.

Campbell, L. and W. Garnett (1882). *The Life of James Clerk Maxwell* (1st edn). London: Macmillan.

Campbell, L. and W. Garnett (1884). *The Life of James Clerk Maxwell* (2nd edn). London: Macmillan.

Carnap, R. (1945). The two conceptions of probability. *Philosophy and Phenomenological Research 5*, 513–532.

Carnap, R. (1950). *The Logical Foundations of Probability.* Chicago: The University of Chicago Press.

Carnot, S. (1824). *Réflexions sur la Puissance Motrice du Feu et sur les Machines Propres a Développer Cette Puissance.* Paris: Bachelier. Translated in Carnot (1890) and Magie (1899).

Carnot, S. (1890). *Reflections on the Motive Power of Fire, and on Machines Fitted to Develop That Power.* London: Macmillan. Translation of Carnot (1824).

Carroll, S. (2010). *From Eternity to Here.* New York: Dutton.

Case, W. C. (2008). Wigner functions and Weyl transforms for pedestrians. *American Journal of Physics 76*, 937–946.

Clausius, R. (1850). Ueber die bewegende Kraft der Wärme und die Gesetze, welche sich daraus für die Wärmelehre selbst ableiten lassen. *Annalen der Physik 79*, 368–397, 500–524. Reprinted in Clausius (1864, 16–78).

Clausius, R. (1854). Ueber eine veränderte Form des zweiten Hauptsatzes der mechanischen Wärmetheorie. *Annalen der Physik 93*, 481–506. Reprinted in Clausius (1864, 127–154); English translation in Clausius (1856) and (1867b).

Clausius, R. (1856). On a modified form of the second fundamental theorem in the mechanical theory of heat. *The London, Edinburgh, and Dublin Philosophical Magazine and Journal of Science 12*, 81–88. English translation of Clausius (1854).

Clausius, R. (1864). *Abhandlungen über die mechanische Wärmetheorie*, vol. 1. Braunschweig: Friedrich Vieweg.

Clausius, R. (1865). Ueber verschiedene für die Anwendung bequeme Formen der Hauptgleichungen der mechanischen Wärmetheorie. *Annalen der Physik 125*, 353–400. Reprinted in Clausius (1867a, 1–44). English translation in Clausius (1867b).

Clausius R. (1867a). *Abhandlungen über die mechanische Wärmetheorie*, vol. 2. Braunschweig: Friedrich Vieweg.

Clausius, R. (1867b). *The Mechanical Theory of Heat, with its Applications to the Steam Engine, and to the Physical Properties of Bodies*. London: John van Voorst. English translation, with one additional paper, of Clausius (1864).

Clausius, R. (1876). *Die Mechanische Wärmetheorie*. Braunschweig: Friedrich Vieweg.

Clausius, R. (1877). Ueber eine von Hrn. Tait in der mechanischen Wärmetheorie angewandte Schlussweise. *Annalen der Physik und Chemie 238*, 130–133.

Clausius, R. (1879). *The Mechanical Theory of Heat*. London: Macmillan. Trans. W. R. Browne. Translation of Clausius (1876).

Clausius, R. (1899). On the motive power of heat, and on the laws that can be deduced from it for the theory of heat. In Magie (1899).

Cournot, A. A. (1843). *Exposition de la Théorie des Chances et des Probabilités*. Paris: L. Hachette.

Crofton, M. W. (1867). Note on local probability. *Mathematical Questions, with their Solutions, from the "Educational Times" 7*, 84–86.

Crofton, M. W. (1868). On the theory of local probability, applied to straight lines drawn at random in a plane, the methods used being also extended to the proof of certain new theorems in the integral calculus. *Philosophical Transactions of the Royal Society of London 158*, 181–199.

d'Alembert, J. (1754). Croix ou pile. In *Encyclopédie ou Dictionnaire raisonné des sciences, des arts et des métiers*, vol. 4, 512–513. Paris: Briasson, David, La Breton, & Durand.

Darrigol, O. (2018). *Atoms, Mechanics, and Probability*. Oxford: Oxford University Press.

de Finetti, B. (1931). *Probabilismo: Saggio critica sulla teoria delle probabilità e sul valore della scienza*. Naples: Francesco Perrella. English translation in de Finetti (1989).

de Finetti, B. (1937). La prevision: ses lois logiques, ses sources subjectives. *Annales de l'Institut Henri Poincaré 7*, 1–68. English translation in de Finetti (1980).

de Finetti, B. (1976). Probability: Beware of falsifications! *Scientia 70*, 283–303. Reprinted in Kyburg and Smokler (1980, 195–224).

de Finetti, B. (1980). Foresight: its logical laws, its subjective sources. In Kyburg and Smokler (1980, 57–118).

de Finetti, B. (1989). Probabilism: A critical essay on the theory of probability and on the value of science. *Erkenntnis 31*, 161–223. Translation, by Maria Concetta Di Maio, Maria Carla Galavotti, and Richard C. Jeffrey, of de Finetti (1931).

de Morgan, A. (1838). *An Essay on Probabilities, and on their Application to Life Contingencies and Insurance Offices.* London: Longman, Brown, Green.

de Morgan, A. (1865). On infinity; and on the sign of equality. *Transactions of the Cambridge Philosophical Society 11*, 145–189.

de Morgan, G. (1866). A solution of the problem of determining the probability that four points taken at random in a plane shall form a re-entrant quadrilateral. *Mathematical Questions, with their Solutions, from the "Educational Times" 5*, 109.

del Rio, L., L. Krämer, and R. Renner (2015). Resource theories of knowledge. arXiv:1511.08818 [quant-ph].

Deutsch, D. (1999). Quantum theory of probability and decisions. *Proceedings of the Royal Society of London A455*, 3129–3137.

Diaconis, P., S. Holmes, and R. Montgomery (2007). Dynamical bias in the coin toss. *SIAM Review 49*, 211–235.

Dirac, P. A. M. (1935). *Principles of Quantum Mechanics* (2nd edn). Oxford: Oxford University Press.

Dove, C. and E. J. Squires (1996). A local model of explicit wavefunction collapse. arXiv:quant-ph/9605047v1.

Dove, C. J. (1996). *Explicit wavefunction collapse and quantum measurement.* PhD thesis, Durham University.

Dürr, D., S. Goldstein, and N. Zanghì (1992). Quantum equilibrium and the origin of absolute uncertainty. *Journal of Statistical Physics 67*, 843–907.

Earman, J. (2006). The "Past Hypothesis": Not even false. *Studies in History and Philosophy of Modern Physics 37*, 399–430.

Easwaren, K. (2016). Conditional probability. In Hájek and Hitchcock (2016, 167–182).

Eddington, A. S. (1931). The end of the world: from the standpoint of mathematical physics. *Nature 127*, 447–453.

Efthymiopoulos, C., G. Contopoulos, and A. Tzemos (2017). Chaos in de Broglie–Bohm quantum mechanics and the dynamics of quantum relaxation. *Annales de la Fondation Louis de Broglie 42*, 133–160.

Ehrenfest, P. and T. Ehrenfest (1912). *Begriffliche Grundlagen der statistischen Auffassung in der Mechanik.* Leipzig: Teubner. English translation in Ehrenfest and Ehrenfest (1959).

Ehrenfest, P. and T. Ehrenfest (1959). *The Conceptual Foundations of the Statistical Approach in Mechanics.* Ithaca, NY: Cornell University Press. Translation, by M. Moravcsik, of Ehrenfest and Ehrenfest (1912).

Einstein, A. (1902). Kinetische Theorie des Wärmegleichgewicht und des zweiten Hauptsätzes der Thermodynamik. *Annalen der Physik 9*, 417–433. Reprinted in Einstein (1989a, 57–75). English translation in Einstein (1989c).

Einstein, A. (1903). Eine Theorie der Grundlagen der Thermodynamik. *Annalen der Physik 11*, 170–187. Reprinted in Einstein (1989a, 77–97). English translation in Einstein (1989d).

Einstein, A. (1936). Physik und Realität. *Journal of the Franklin Institute 221*, 349–382. English translation in Einstein (1954).

Einstein, A. (1954). Physics and reality. In *Ideas and Opinions*, 290–323. New York: Crown. Translation of Einstein (1936).

Einstein, A. (1989a). *The Collected Papers of Albert Einstein, vol. 2: The Swiss Years: Writings, 1900–1909*. Princeton, NJ: Princeton University Press.

Einstein, A. (1989b). *The Collected Papers of Albert Einstein, vol. 2: The Swiss Years: Writings, 1900–1909* (English translation supplement). Princeton, NJ: Princeton University Press.

Einstein, A. (1989c). Kinetic theory of thermal equilibrium and of the second law of thermodynamics. In Einstein (1989b, 30–47).

Einstein, A. (1989d). A theory of the foundations of thermodynamics. In Einstein (1989b, 49–67).

Ellis, R. L. (1849). On the foundations of the theory of probabilities. *Transactions of the Cambridge Philosophical Society 8*, 1–6. Read before the Society Feb. 14, 1842.

Engel, E. M. (1992). *A Road to Randomness in Physical Systems*. Berlin: Springer.

Esseen, C. G. (1956). A moment inequality with an application to the central limit theorem. *Skandinavisk Aktuarietidskrift 39*, 160–170.

Farmer, J. S. and W. E. Henley (1891). *Slang and its Analogues Past and Present: A Dictionary, Historical and Comparative, of the Heterodox Speech of All Classes of Society for More than Three Hundred Years*, vol. 2: *C–Fizzle*. London: Printed for Subscribers Only.

Feller, W. (1957). *An Introduction to Probability Theory and Its Applications* (2nd edn), vol. 1. New York: John Wiley.

Feller, W. (1970). *An Introduction to Probability Theory and Its Applications* (2nd edn), vol. 2. New York: John Wiley.

Fisher, R. A. (1950). *Contributions to Mathematical Statistics*. New York: John Wiley.

Fiske, T. S. (1894). The summer meeting of the American Mathematical Society. *Bulletin of the American Mathematical Society 1*, 1–6.

French, S. and J. Saatsi (eds) (2020). *Scientific Realism and the Quantum*. Oxford: Oxford University Press.

Frigg, R., B. J. and F. Kronz (2020). The ergodic hierarchy. In E. N. Zalta (ed.), *The Stanford Encyclopedia of Philosophy* (Fall 2020). Stanford, CA: Metaphysics Research Lab, Stanford University.

Frigg, R. (2010). Probability in Boltzmannian statistical mechanics. In G. Ernst and A. Hütteman (eds), *Time, Chance, And Reduction: Philosophical Aspects of Statistical Mechanics*, 92–118. Cambridge: Cambridge University Press.

Fuchs, C. A. and B. Stacey (2019). QBism: Quantum theory as a hero's handbook. In E. M. Rasel, W. P. Schleich, and S. Wölk (eds), *Foundations of Quantum Theory: Proceedings of the International School of Physics "Enrico Fermi" Course 197*, 133–202. Amsterdam: IOS Press.

Galton, F. (1885). The application of a graphic method to fallible measures [with discussion]. *Journal of the Statistical Society of London Jubilee Volume*, 262–271.

Gao, S. (ed.) (2018). *Collapse of the Wave Function: Models, Ontology, Origin, and Implications*. Cambridge: Cambridge University Press.

Garber, E., S. G. Brush, and C. W. F. Everitt (eds) (1995). *Maxwell on Heat and Statistical Mechanics: On "Avoiding All Personal Enquiries" of Molecules*. Bethlehem, PA: Lehigh University Press.

Ghirardi, G. C., P. Pearle, and A. Rimini (1990). Markov processes in Hilbert space and continuous spontaneous localization of systems of identical particles. *Physical Review A 42*, 78–89.

Ghirardi, G. C., A. Rimini, and T. Weber (1986). Unified dynamics for microscopic and macroscopic systems. *Physical Review D 34*, 470–491.

Gibbs, J. W. (1875, 1877). On the equilibrium of heterogeneous substances. *Transactions of the Connecticut Academy of Arts and Sciences 3*, 108–248, 343–524. Reprinted in Gibbs (1906a, 55–353).

Gibbs, J. W. (1885). On the fundamental formula of statistical mechanics, with applications to astronomy and thermodynamics. *Proceedings of the American Association for the Advancement of Science 33*, 57–58. Reprinted in Gibbs (1906b, p. 16).

Gibbs, J. W. (1902). *Elementary Principles in Statistical Mechanics: Developed with Especial Reference to the Rational Foundation of Thermodynamics*. New York: Charles Scribner's Sons.

Gibbs, J. W. (1906a). *The Scientific Papers of J. Willard Gibbs, PhD, LLD*, vol. 1. New York: Longmans, Green.

Gibbs, J. W. (1906b). *The Scientific Papers of J. Willard Gibbs, PhD, LLD*, vol. 2. New York: Longmans, Green.

Godfray, H. (1866). On some problems in the theory of chances. *Mathematical Questions, with their Solutions, from the "Educational Times" 6*, 72–74.

Godfray, H. (1867). At random. *Mathematical Questions, with their Solutions, from the "Educational Times" 7*, 65–67.

Goldstein, S. (2001). Boltzmann's approach to statistical mechanics. In J. Bricmont, D. Dürr, M. Galavotti, G. Ghirardi, F. Petruccione, and N. Zanghì (eds), *Chance in Physics*, 39–54. Berlin: Springer.

Good, I. J. (1952). Rational decisions. *Journal of the Royal Statistical Society Series B 14*, 107–114. Reprinted in Good (1983, 3–14).

Good, I. J. (1967). On the principle of total evidence. *The British Journal for the Philosophy of Science 17*, 319–322.

Good, I. J. (1968). The white shoe *qua* herring is pink. *The British Journal for the Philosophy of Science 19*, 156–157.

Good, I. J. (1971). 46656 varieties of Bayesians. *The American Statistician 25*, 62–63. Reprinted in Good (1983, 20–21).

Good, I. J. (1979). Some history of the hierarchical Bayesian methodology. In J. M. Bernardo (ed.), *Bayesian Statistics*, 489–515. Valencia: University of Valencia Press. Reprinted in Good (1983, 95–105).

Good, I. J. (1983). *Good Thinking: The Foundations of Probability and its Applications*. Minneapolis: The University of Minnesota Press.

Goold, J., Huber, M., Riera, A., del Rio, L., and Skrzypczyk, P. (2016). The role of quantum information in thermodynamics a topical review. *Journal of Physics A: Mathematical and Theoretical*, 49:143001.

Gour, G., M. P. Müller, V. Narasimhachar, R. W. Spekkens, and N. Y. Halpern (2015). The resource theory of informational nonequilibrium in thermodynamics. *Physics Reports 583*, 1–58.

Grattan-Guinness, I. (1992). A note on *The Educational Times* and *Mathematical Questions*. *Historia Mathematica 19*, 76–78.

Greaves, H. (2007). On the Everettian epistemic problem. *Studies in History and Philosophy of Modern Physics 38*, 120–152.

Greaves, H. and W. C. Myrvold (2010). Everett and evidence. In Saunders et al. (2010, 264–304).

Hacking, I. (1971). Equipossibility theories of probability. *The British Journal for the Philosophy of Science 22*, 339–355.

Hacking, I. (1975). *The Emergence of Probability*. Cambridge: Cambridge University Press.

Hacking, I. (1990). *The Taming of Chance*. Cambridge: Cambridge University Press.

Hahn, E. L. (1950). Spin echoes. *Physical Review 80*, 580–594.

Hahn, E. L. (1953). Free nuclear induction. *Physics Today 6*, 4–9.

Hájek, A. (1997). "Mises redux"—redux: Fifteen arguments against finite frequentism. *Erkenntnis 45*, 209–227.

Hájek, A. (2003). What conditional probability could not be. *Synthese 137*, 273–323.

Hájek, A. (2009). Fifteen arguments against hypothetical frequentism. *Erkenntnis 70*, 211–235.

Hájek, A. (MS). Staying regular? Available at http://hplms.berkeley.edu/HajekStayingRegular.pdf.

Hájek, A. and C. Hitchcock (eds) (2016). *The Oxford Handbook of Probability and Philosophy*. Oxford: Oxford University Press.

Harman, P. M. (ed.) (1990). *The Scientific Letters and Papers of James Clerk Maxwell*, vol. 1: *1846–1862*. Cambridge: Cambridge University Press.

Harman, P. M. (ed.) (1995). *The Scientific Letters and Papers of James Clerk Maxwell*, vol. 2: *1862–1873*. Cambridge: Cambridge University Press.

Harman, P. M. (ed.) (2002). *The Scientific Letters and Papers of James Clerk Maxwell*, vol. 3: *1874–1879*. Cambridge: Cambridge University Press.

Healey, R. (2017a). Quantum states as objective informational bridges. *Foundations of Physics 47*, 161–173.

Healey, R. (2017b). *The Quantum Revolution in Philosophy*. Oxford: Oxford University Press.

Healey, R. (2020). Pragmatist quantum realism. In French and Saatsi (2020, 123–146).

Hilbert, S., P. Hänggi, and J. Dunkel (2014). Thermodynamic laws in isolated systems. *Physical Review E 90*, 062116.

Hillery, M., R. F. O'Connell, M. O. Scully, and E. P. Wigner (1984). Distribution functions in physics: Fundamentals. *Physics Reports 106*, 121–167.

Hoefer, C. (2016). Objective chance: not propensity, maybe determinism. *Lato Sensu: Revue de la Société de philosophie des sciences 3*, 31–42.

Hopf, E. (1934). On causality, statistics, and probability. *Journal of Mathematics and Physics 13*, 51–102.

Hopf, E. (1936). Über die Bedeutung der willkürlichen Funktionen für die Wahrscheinlichkeitstheorie. *Jahresbericht des deutschen Mathematiker-Vereinigung 46*, 179–195.

Hurwicz, L. (1951). Some specification problems and applications to econometric models. *Econometrica 19*, 343–344.

Ingleby, C. M. (1866a). On a problem in the theory of chances. *Mathematical Questions, with their Solutions, from the "Educational Times" 5*, 81–82.

Ingleby, C. M. (1866b). Correction of an inaccuracy in Dr. Ingleby's note on the four-point problem. *Mathematical Questions, with their Solutions, from the "Educational Times" 5*, 108–109.

Ismael, J. (2009). Probability in deterministic physics. *Journal of Philosophy 106*, 89–108.

Ismael, J. (2015). Quantum mechanics. In E. N. Zalta (ed.), *The Stanford Encyclopedia of Philosophy* (Spring 2015). Stanford, CA: Metaphysics Research Lab, Stanford University.

Ismael, J. T. (2007). *The Situated Self*. Oxford: Oxford University Press.

Ismael, J. T. (2016). *How Physics Makes Us Free*. Oxford: Oxford University Press.

Jackson, E. A. (1968). *Equilibrium Statistical Mechanics*. New York: Dover.

Jaynes, E. (1957a). Information theory and statistical mechanics, I. *Physical Review 106*, 620–630. Reprinted in Jaynes (1989a, 7–16).

Jaynes, E. (1957b). Information theory and statistical mechanics, II. *Physical Review 108*, 171–190. Reprinted in Jaynes (1989a, 19–37).

Jaynes, E. (1989a). *Papers on Probability, Statistics, and Statistical Physics*. Dordrecht: Kluwer Academic.

Jaynes, E. (1989b). Where do we stand on maximum entropy? In Jaynes (1989a, 211–314).

Jaynes, E. T. (1973). The well posed problem. *Foundations of Physics 3*, 477–493. Reprinted in Jaynes (1989a, 133–148).

Jaynes, E. T. (2003). *Probability Theory: The Logic of Science*. Cambridge: Cambridge University Press.

Jeffrey, R. (1992). Mises redux. In *Probability and the Art of Judgment*, 192–202. Cambridge: Cambridge University Press.

Jevons, W. S. (1874). *The Principles of Science: A Treatise on Logic and Scientific Method*. London: Macmillan.

Joyce, J. M. (1998). A nonpragmatic vindication of probabilism. *Philosophy of Science 65*, 573–603.

Joyce, J. M. (2009). Accuracy and coherence: Prospects for an alethic epistemology of partial belief. In F. Huber and C. Schmidt-Petri (eds), *Degrees of Belief*, 263–297. Berlin: Springer.

Kadanoff, L. (2000). *Statistical Physics: Statics, Dynamics, and Renormalization.* Singapore: World Scientific.

Keynes, J. M. (1921). *A Treatise on Probability.* London: Macmillan.

Khintchine, A. (1929). Sur la loi des grand nombres. *Comptes Rendus Hebdomadaires des Séances de l'Académie de Sciences 188*, 477–479.

Knott, C. G. (1911). *Life and Scientific Work of Peter Guthrie Tait.* Cambridge: Cambridge University Press.

Kolmogorov, A. N. (1933). *Grundbegriffe der Wahrscheinlichkeitsrechnung.* Berlin: Julius Springer.

Kolmogorov, A. N. and S. Fomin (1975). *Introductory Real Analysis.* New York: Dover. Trans. and ed. R. A. Silverman.

Kožnjak, B. (2015). Who let the demon out? Laplace and Boscovich on determinism. *Studies in History and Philosophy of Science 51*, 42–52.

Krönig, A. (1856). Grundzüge einer Theorie der Gas. *Annalen der Physik 175*, 315–322.

Kyburg, H. E. and H. E. Smokler (eds) (1980). *Studies in Subjective Probability.* Huntington, NY: Robert E. Krieger.

Laplace, P.-S. (1776). Recherches sur l'integration des équations différentielles aux différence finies, et leur usage dans la théorie des hasards. *Mémoires de Mathématique et de Physique Présentés à l'Académie Royale des Sciences 7*, 37–232. Reprinted in *Oeuvres Complètes de Laplace*, vol. 8, 69–196.

Laplace, P.-S. (1814). *Essai Philosophique sur les Probabilités.* Paris: Courcier. English translation in Laplace (1902).

Laplace, P.-S. (1902). *A Philosophical Essay on Probabilities.* New York: John Wiley. Translation of Laplace (1814).

Lavis, D. A. (2004). The spin-echo system reconsidered. *Foundations of Physics 34*, 669–688.

Lebowitz, J. L. (1999a). Microscopic origins of irreversible macroscopic behavior. *Physica A 263*, 516–527.

Lebowitz, J. L. (1999b). Statistical mechanics: A selective review of two central issues. *Reviews of Modern Physics 71*, S346–S357.

Lee, H.-W. (1995). Theory and application of the quantum phase-space distribution functions. *Physics Reports 259*, 147–211.

Lehman, R. S. (1955). On confirmation and rational betting. *The Journal of Symbolic Logic 20*, 251–262.

Lewis, D. (1980). A subjectivist's guide to objective chance. In R. C. Jeffrey (ed.), *Studies in Inductive Logic and Probability*, vol. 2, 263–93. Berkeley: University of California Press.

Lewis, D. (1994). Humean supervenience debugged. *Mind 103*, 473–490.

Linden, N., S. Popescu, A. J. Short, and A. Winter (2009). Quantum mechanical evolution towards thermal equilibrium. *Physical Review E 79*, 061103.

Loewer, B. (2001). Determinism and chance. *Studies in History and Philosophy of Modern Physics 32*, 609–620.

London, F. and E. Bauer (1939). *La Théorie de l'Observation en Méchanique Quantique.* Paris: Hermann. English translation in London and Bauer (1983).

London, F. and E. Bauer (1983). The theory of observation in quantum mechanics. In Wheeler and Zurek (1983, 217–259).

Loschmidt, J. (1876). Über den Zustand des Wärmegleichwicht eines systems von Körpen mit Rücksicht auf die Schwerkraft. *Sitzungberichte der Akademie der Wissenschaften zu Wien, mathematisch-naturwissenschaftliche Klasse 73*, 128–142.

Lostaglio, M. (2019). An introductory review of the resource theory approach to thermodynamics. *Reports in Progress in Physics*, 82:114001.

Magie, W. F. (ed.) (1899). *The Second Law of Thermodynamics: Memoirs by Carnot, Clausius, and Thomson*. New York: Harper.

Maroney, O. (2007). The physical basis of the Gibbs–von Neumann entropy. arXiv: quant-ph/0701127v2.

Maxwell, J. C. (1871). *Theory of Heat*. London: Longmans, Green.

Maxwell, J. C. (1873a). Does the progress of physical science tend to give any advantage to the opinion of necessity (or determinism) over that of the contingency of events and the freedom of the will? Published in Campbell and Garnett (1882, 434–444; 1884, 357–366), and in Harman (1995, 814–823).

Maxwell, J. C. (1873b). Molecules. *Nature 8*, 437–441. Reprinted in Niven (1890, 361–377).

Maxwell, J. C. (1877). Diffusion. In *Encyclopaedia Britannica* (9th edn), vol. 7, 214–221. Edinburgh: Adam and Charles Black. Reprinted in Niven (1890, 625–646).

Maxwell, J. C. (1878). Tait's "Thermodynamics". *Nature 17*, 257–259, 278–280. Reprinted in Niven (1890, 660–671).

Miller, D. (1966). A paradox of information. *The British Journal for the Philosophy of Science 17*, 59–61.

Miller, J. (n.d.). Earliest known uses of some of the words of mathematics. http://jeff560.tripod.com/mathword.html.

Miller, W. J. (1857). Quest. (1904), Second solution. *The Lady's and Gentleman's Diary 154*, 55–56.

Miller, W. J. C. (1878). Notes on random chords. *Mathematical Questions, with their Solutions, from the "Educational Times" 29*, 17–20.

Myrvold, W. C. (2005). Why I am not an Everettian. users.ox.ac.uk/~everett/docs/Myrvold%20Not%20an%20Everettian.pdf.

Myrvold, W. C. (2012a). Deterministic laws and epistemic chances. In Y. Ben-Menahem and M. Hemmo (eds), *Probability in Physics*, 73–85. Berlin: Springer.

Myrvold, W. C. (2012b). Epistemic values and the value of learning. *Synthese 87*, 547–568.

Myrvold, W. C. (2015). You can't always get what you want: Some considerations concerning conditional probability. *Erkenntnis 80*, 573–603.

Myrvold, W. C. (2016). Lessons of Bell's theorem: Nonlocality, yes; action at a distance, not necessarily. In Bell and Gao (2016, 238–260).

Myrvold, W. C. (2018a). Ontology for collapse theories. In Gao (2018, 97–123).

Myrvold, W. C. (2018b). Philosophical issues in quantum theory. In E. N. Zalta (ed.), *The Stanford Encyclopedia of Philosophy* (Fall 2018). Stanford, CA: Metaphysics Research Lab, Stanford University.

Myrvold, W. C. (2019a). Learning is a risky business. *Erkenntnis 84*, 477–584.

Myrvold, W. C. (2019b). Ontology for relativistic collapse theories. In O. Lombardi, S. Fortin, C. López, and F. Holik (eds.), *Quantum Worlds: Perspectives on the Ontology of Quantum Mechanics*. Cambridge: Cambridge University Press 9–31.

Myrvold, W. C. (2020a). Explaining thermodynamics: What remains to be done? In V. Allori (ed.), *Statistical Mechanics and Scientific Explanation: Determinism, Indeterminism and Laws of Nature*. Singapore: World Scientific, 113–143.

Myrvold, W. C. (2020b). On the status of quantum state realism. In French and Saatsi (2020, 219–251).

Myrvold, W. C. (2020c). Subjectivists about quantum probabilities should be realists about quantum states. In M. Hemmo and O. Shenker (eds), *Quantum, Probability, Logic: The Work and Influence of Itamar Pitowsky*. Berlin: Springer Scientific, 449–465.

Myrvold, W. C. (2020d). The science of $\Theta\Delta^{cs}$. *Foundations of Physics 50*, 1219–1251.

Nagel, E. (1955). *Fundamentals of the Theory of Probability*. Chicago: Chicago University Press.

Niven, W. D. (ed.) (1890). *The Scientific Papers of James Clerk Maxwell*, vol. 2. Cambridge: Cambridge University Press.

Norton, J. (2016). The impossible process: Thermodynamic reversibility. *Studies in History and Philosophy of Modern Physics 55*, 43–61.

Ore, O. (1960). Pascal and the invention of probability theory. *American Mathematical Monthly 67*, 409–419.

Pearle, P. (1989). Combining stochastic dynamical state-vector reduction with spontaneous localization. *Physical Review A 39*, 913–923.

Pearle, P. (2015). Relativistic dynamical collapse model. *Physical Review D 91*, 105012.

Penrose, O. (1970). *Foundations of Statistical Mechanics: A Deductive Approach*. Oxford: Pergamon Press.

Penrose, R. (1979). Singularities and time-asymmetry. In S. W. Hawking and W. Israel (eds), *General Relativity: An Einstein Centenary Survey*, 581–638. Cambridge: Cambridge University Press.

Penrose, R. (1989a). *The Emperor's New Mind*. Oxford: Oxford University Press.

Penrose, R. (1989b). Difficulties with inflationary cosmology. *Proceedings of the New York Academy of Science 571*, 249–264.

Pettigrew, R. (2016a). Accuracy, risk, and the principle of indifference. *Philosophy and Phenomenological Research 92*, 35–59.

Pettigrew, R. (2016b). *Accuracy and the Laws of Credence*. Oxford: Oxford University Press.

Poincaré, H. (1890). Sur le problème des trois corps et les équations de dynamique. *Acta Mathematica 13*, 8–270.

Poincaré, H. (1896). *Calcul des probabilités*. Paris: Gauthier-Villars.

Poincaré, H. (1899). Réflexions sur le calcul des probabilités. *Revue générale des sciences pures et appliquées 10*, 262–269.

Poincaré, H. (1902). *La Science et l'Hypothèse*. Paris: Flammarion. English translation in Poincaré (2018).

Poincaré, H. (1908). *Science et Méthode*. Paris: Flammarion.

Poincaré, H. (1912). *Calcul des probabilités* (2nd edn). Paris: Gauthier-Villars.

Poincaré, H. (1966). On the three-body problem and the equations of dynamics. In Brush (1966, 368–376). English translations of parts of Poincaré (1890).

Poincaré, H. (2018). *Science and Hypothesis*. London: Bloomsbury Academic. Translation, by M. Frappier, A. Smith, and D. J. Stump, of Poincaré (1902).

Poisson, S.-D. (1835). Calcul des probabilités—recherches sur la probabilité des jugements, principalement en matière criminelle. *Comptes Rendus Hebdomadaires des Séances de l'Académie des Sciences 1*, 473–494.

Poisson, S.-D. (1837). *Recherches sur la Probabilité des Jugements en Matière Criminelle et en Matière Civile, précédées des règles générales du calcul des probabilités*. Paris: Bachelier.

Popper, K. R. (1957). The propensity interpretation of the calculus of probability, and the quantum theory. In S. Körner (ed.), *Observation and Interpretation: A Symposium of Philosophers and Physicists*, 65–70. London: Butterworths.

Popper, K. R. (1959). The propensity interpretation of probability. *The British Journal for the Philosophy of Science 37*, 25–42.

Popper, K. R. (1982). *Quantum Theory and the Schism in Physics*. Totowa, NJ: Rowman & Littlefield.

Porter, T. M. (1986). *The Rise of Statistical Thinking 1820–1900*. Princeton, NJ: Princeton University Press.

Price, H. (2002). Boltzmann's time bomb. *The British Journal for the Philosophy of Science 53*, 83–119.

Pusey, M. A., J. Barrett, and T. Rudolph (2012). On the reality of the quantum state. *Nature Physics 8*, 475–478.

Pusz, W. and S. L. Woronowicz (1978). Passive states and KMS states for general quantum systems. *Communications in Mathematical Physics 58*, 273–290.

Quetelet, A. (1835). *Sur l'Homme et le Développement de ses Facultés, ou, Essai de Physique Sociale*. Paris: Bachelier.

Ramsey, F. P. (1931). Truth and probability. In *The Foundations of Mathematics and other Logical Essays*, 156–198. London: Routledge & Kegan Paul.

Rankine, W. J. M. (1859). *A Manual of the Steam Engine and Other Prime Movers*. London: Richard Griffin.

Rawson, R. W. (1885). Presidential address. *Journal of the Statistical Society of London Jubilee Volume*, 1–13.

Rees, M. (1997). *Before the Beginning: Our Universe and Others*. Boston, MA: Addison-Wesley.

Reif, F. (1968). *Fundamentals of Statistical and Thermal Physics*. New York: McGrawHill.

Ridderbos, K. (2002). The coarse-graining approach to statistical mechanics: how blissful is our ignorance? *Studies in History and Philosophy of Modern Physics 33*, 65–77.

Ridderbos, T. M. and M. L. G. Redhead (1998). The spin-echo experiments and the second law of thermodynamics. *Foundations of Physics 28*, 1237–1270.

Rosenthal, J. (2012). Probabilities as ratios of ranges in initial-state spaces. *Journal of Logic, Language, and Information 21*, 217–236.

Rosenthal, J. S. (2006). *A First Look at Rigorous Probability Theory* (2nd edn). Singapore: World Scientific.

Rutherford, Dr. (1857). Quest. (1904), Solution. *The Lady's and Gentleman's Diary 154*, 55.

Saunders, S., J. Barrett, A. Kent, and D. Wallace (eds) (2010). *Many Worlds? Everett, Quantum Theory, and Reality*. Oxford: Oxford University Press.

Savage, L. J. (1954). *The Foundations of Statistics*. New York: John Wiley.

Savage, L. J. (1973). Probability in science: A personalistic account. In P. Suppes (ed.), *Logic Methodology, and Philosophy of Science IV*, 417–428. Amsterdam: North-Holland.

Schaffer, J. (2007). Deterministic chance? *The British Journal for the Philosophy of Science 58*, 113–140.

Schrödinger, E. (1944). *What Is Life? The Physical Aspect of the Living Cell*. Cambridge: Cambridge University Press. Reprinted in Schrödinger (1967).

Schrödinger, E. (1967). *What Is Life? with Mind and Matter and Autobiographical Sketches*. Cambridge: Cambridge University Press.

Sebens, C. T. and S. M. Carroll (2018). Self-locating uncertainty and the origin of probability in Everettian quantum mechanics. *The British Journal for the Philosophy of Science 69*, 25–74.

Seidenfeld, T., M. J. Schervish, and J. B. Kadane (1995). A representation of partially ordered preferences. *The Annals of Statistics 23*, 2168–2217.

Shevtsova, I. G. (2013). On the absolute constants in the Berry–Esseen inequality and its structural and nonuniform improvements. *Informatika i Ee Primeneniya 7*, 124–125.

Shimony, A. (1971). Scientific inference. In R. Colodny (ed.), *The Nature and Function of Scientific Theories*, 79–172. Pittsburgh, PA: Pittsburgh University Press. Reprinted in Shimony (1993).

Shimony, A. (1993). *Seach for a Naturalistic World View*, vol. 1: *Scientific Method and Epistemology*. Cambridge: Cambridge University Press.

Sklar, L. (1993). *Physics and Chance*. Cambridge: Cambridge University Press.

Solovay, R. M. (1970). A model of set-theory in which every set of reals is Lebesgue measurable. *Annals of Mathematics 92*, 1–56.

Statistical Society of London (1838). Introduction. *Journal of the Statistical Society of London 1*, 1–5.

Stigler, S. M. (1999). *Statistics on the Table*. Cambridge, MA: Harvard University Press.

Strevens, M. (2003). *Bigger Than Chaos: Understanding Compexity Through Probability*. Cambridge, MA: Harvard University Press.

Strevens, M. (2011). Probability out of determinism. In Beisbart and Hartmann (2011, 339–364).

Strevens, M. (2013). *Tychomancy: Inferring Probability from Causal Structure*. Cambridge, MA: Harvard University Press.

Szilard, L. (1925). Über die Ausdehnung der phänomenologischen Thermodynamik auf die Schwankungserscheinungen. *Zeitschrift für Physik 32*, 753–788. English translation in Szilard (1972).

Szilard, L. (1972). On the extension of phenomenological thermodynamics to fluctuation phenomena. In B. T. Feld, G. W. Szilard, and K. R. Winsor (eds), *The Collected Works of Leo Szilard: Scientific Papers*, 70–102. Cambridge, MA: MIT Press.

Tait, P. G. (1868). *Sketch of Thermodynamics*. Edinburgh: Edmonston & Douglas.

Tait, P. G. (1876). *Lectures on Some Recent Advances in Physical Science* (2nd edn). London: Macmillan.

Tait, P. G. (1877). *Sketch of Thermodynamics* (2nd edn). Edinburgh: David Douglas.

Thomson, H. (1878). On the random chord question. *Mathematical Questions, with their Solutions, from the "Educational Times"* 29, 40.

Thomson, W. (1848). On an absolute thermometric scale founded on Carnot's theory of the motive power of heat, and calculated from Regnault's observations. *Proceedings of the Cambridge Philosophical Society* 1(5), 66–71. Reprinted in Thomson (1882, 100–106).

Thomson, W. (1853). On the dynamical theory of heat, with numerical results deduced from Mr Joule's equivalent of a thermal unit, and M. Regnault's observations on steam. *Transactions of the Royal Society of Edinburgh* 20, 261–288. Reprinted in Thomson (1882, 174–210).

Thomson, W. (1857). On the dynamical theory of heat. Part VI: Thermo-electric currents. *Transactions of the Royal Society of Edinburgh* 21, 123–171. Reprinted in Thomson (1882, 232–291).

Thomson, W. (1874). The kinetic theory of the dissipation of energy. *Nature* 9, 441–444. Reprinted in Thomson (1911, 11–20).

Thomson, W. (1882). *Mathematical and Physical Papers*, vol. 1. Cambridge: Cambridge University Press.

Thomson, W. (1911). *Mathematical and Physical Papers*, vol. 5. Cambridge: Cambridge University Press.

Tolman, R. C. (1938). *The Principles of Statistical Mechanics*. Oxford: Clarendon Press.

Tumulka, R. (2006). A relativistic version of the Ghirardi–Rimini–Weber model. *Journal of Statistical Physics* 125, 825–844.

Uffink, J. (2006). Insuperable difficulties: Einstein's statistical road to molecular physics. *Studies in History and Philosophy of Modern Physics* 37, 36–70.

Uffink, J. (2011). Subjective probability and statistical physics. In Beisbart and Hartmann (2011, 25–49).

Uffink, J. (2017). Boltzmann's work in statistical physics. In E. N. Zalta (ed.), *The Stanford Encyclopedia of Philosophy* (Spring 2017 edn). Stanford, CA: Metaphysics Research Lab, Stanford University.

Uhlenbeck, G. E. and G. W. Ford (1963). *Lectures in Statistical Mechanics*. Providence, RI: American Mathematical Society.

Vaidman, L. (1998). On schizophrenic experiences of the neutron, or why we should believe in the many-worlds interpretation of quantum theory. *International Studies in the Philosophy of Science* 12, 245–261.

Valentini, A. (1991a). Signal-locality, uncertainty, and the sub-quantum H-theorem. I. *Physics Letters A* 156, 5–11.

Valentini, A. (1991b). Signal-locality, uncertainty, and the sub-quantum H-theorem. II. *Physics Letters A 158*, 1–8.

Valentini, A. (2002a). Signal-locality and subquantum information in deterministic hidden-variables theories. In T. Placek and J. Butterfield (eds), *Non-locality and Modality*, 81–104. Berlin: Springer.

Valentini, A. (2002b). Signal-locality in hidden-variables theories. *Physics Letters A 297*, 273–278.

Valentini, A. (2002c). Subquantum information and computation. *Pramana 59*, 269–277.

van Fraassen, B. C. (1977). The pragmatics of explanation. *American Philosophical Quarterly 14*, 143–150.

van Fraassen, B. C. (1980). *The Scientific Image*. Oxford: Oxford University Press.

Venn, J. (1866). *The Logic of Chance*. London: Macmillan.

Venn, J. (1888). *The Logic of Chance* (3rd edn). London: Macmillan.

von Kries, J. (1886). *Die Principien Der Wahrscheinlichkeitsrechnung: Eine Logische Untersuchung*. Freiburg: Mohr.

von Mises, R. (1928). *Wahrscheinlichkeit Statistik und Wahrheit*. Vienna: Springer Verlag.

von Mises, R. (1957). *Probability, Statistics, and Truth*. London: George Allen & Unwin. Second revised English edn, prepared by H. Geiringer.

von Plato, J. (1983). The method of arbitrary functions. *The British Journal for the Philosophy of Science 34*, 37–47.

Wallace, D. (2003). Everettian rationality: Defending Deutsch's approach to probability in the Everett interpretation. *Studies in History and Philosophy of Modern Physics 34*, 415–439.

Wallace, D. (2006). Epistemology quantized: Circumstances in which we should come to believe the Everett interpretation. *British Journal for the Philosophy of Science 57*, 655–689.

Wallace, D. (2007). Quantum probability from subjective likelihood: Improving on Deutsch's proof of the probability rule. *Studies in History and Philosophy of Modern Physics 38*, 311–332.

Wallace, D. (2012). *The Emergent Multiverse*. Oxford: Oxford University Press.

Wallace, D. (2016a). Probability and irreversibility in modern statistical mechanics: Classical and quantum. Forthcoming in D. Bedingham, O. Maroney, and C. Timpson (eds), *Quantum Foundations of Statistical Mechanics* (Oxford University Press). Preprint available at http://dornsife.usc.edu/assets/sites/1045/docs/oxfordstatmech.pdf.

Wallace, D. (2016b). Thermodynamics as control theory. *Entropy 16*, 699–725.

Weinberg, S. (1995). *The Quantum Theory of Fields*, vol. 1. Cambridge: Cambridge University Press.

Wheeler, J. A. and W. H. Zurek (eds) (1983). *Quantum Theory and Measurement*. Princeton, NJ: Princeton University Press.

Whitworth, Rev. Prof. (1866). Note on Dr. Ingleby's strictures on Mr. Wilson's solution of a problem in chances. *Mathematical Questions, with their Solutions, from the "Educational Times" 5*, 109.

Wigner, E. (1932). On the quantum correction for thermodynamic equilibrium. *Physical Review 40*, 749–759.

Wilson, J. M. (1866a). On the four-point and similar geometrical chance problems. *Mathematical Questions, with their Solutions, from the "Educational Times" 5*, 81.

Wilson, J. M. (1866b). On the four-point problem. *Mathematical Questions, with their Solutions, from the "Educational Times" 6*, 82.

Woolhouse, W. S. B. (1866a). Observations on the "three and four point problems" in their relations to infinity. *Mathematical Questions, with their Solutions, from the "Educational Times" 6*, 49–52.

Woolhouse, W. S. B. (1866b). Note on question 1894. *Mathematical Questions, with their Solutions, from the "Educational Times" 6*, 81.

Woolhouse, W. S. B. (1867). Some additional observations on the four-point problem. *Mathematical Questions, with their Solutions, from the "Educational Times" 7*, 81–83.

Woolhouse, W. S. B. (1868). Note on random lines. *Mathematical Questions, with their Solutions, from the "Educational Times" 10*, 33.

Woolhouse, W. S. B. (1877). Note on question 5461. *Mathematical Questions, with their Solutions, from the "Educational Times" 28*, 109–110.

Zermelo, E. (1896). Ueber einen Satz der Dynamik und die mechanische Wärmetheorie. *Annalen der Physik, Neue Folge 57*, 485–494. English translation in Zermelo (1966).

Zermelo, E. (1966). On a theorem of dynamics and the mechanical theory of heat. In Brush (1966, 382–391). English translation of Zermelo (1896).

Index

Dirac, P. A. M. 218, 230–1
  collapse postulate 230
distribution
  attractor 78, 87–92, 106, 179, 217
  canonical 151–3, 157–8, 160–1, 192,
    198, 232
  equilibrium 88–9, 92–96, 151–4, 169–70,
    175–6, 198, 216
  grand canonical 153–4
  invariant 89–92, 98–9, 121, 169–70
  input 83–4, 94–5, 107–10, 118–19, 121
  microcanonical 153, 165, 175–7, 181,
    198, 232
  normal 7, 84, 256–8
    defined 256
  stationary 150, 219
  uniform 55, 56, 81, 93
Dove, C. J. 218
dutch book (argument) 22, 62
  origin of term 22 fn 7
Dürr, D. 212 see also DGZ

Earman, J. 201 fn 12
Easwaren, K. 243
Eddington, A. S. 166 fn 12
Edgeworth, F. Y. 6
effectively random (behaviour, variable,
  motion, etc.) 3, 13–15, 186
Efthymiopoulos, C. 217
Ehrenfest, P. and Ehrenfest, T.
  xv, 162
eigenstate-eigenvalue link 21, 227
Einstein, A. 146, 157 fn 8, 167, 205
Ellis, R. L. 42 fn 17
energy
  as conserved quantity (or conservation
    of) 5, 123–4, 138
  conservation of, violated in collapse
    theories 219
  dissipation of 124–5, 142–3, 150
  internal 126
  variance, for canonical
    distribution 153, 156
Engel, E. M. 98, 107
ensemble 151
entropy
  as means-relative 143–4
  Boltzmann 162–4
  coarse-grained 190–1
  Gibbs 156–7, 167–8, 190
  of mixing 142

non-equilibrium 160–1
origin of term 132 fn 7
principle of maximum 62–4
statistical-mechanical analogs of
  154–1
surface 157
thermodynamic 130–4
volume 157
von Neumann 157–8, 233
epistemic chance
  as hybrid notion xiv, 110, 120, 240
  definition 111–12
  distinguished from credence 110
  learning about 117–17
  statistical mechanical probabilities
    as 180–1
Epistemic Chance Principal Principle
  (ECPP) 114
equilibration
  as convergence of probability
    distributions 179–80
  explanation of 194–7
  partial 95–6, 181–86
equilibrium
  distribution 88–9, 92–6, 151–4, 169–70,
    175–6, 198, 216
  statistical-mechanical 137–8
  thermal 125–6, 184
  thermodynamic 118, 125, 164–5
Equilibrium (or Equilibration)
  Principle 128–30
equivariance 214, 216–17
ergodic theory 178–9
Esseen, C. G. 257
Everitt, C. W. F. 135 fn 9
Everettian interpretations of quantum
  mechanics 212, 220–4
evidence
  accessible 112–14
  admissible 33, 114
  about chance 33, 67
evolution
  Hamiltonian 149
  of density functions 97, 150
  of parabola gadget 78–9
  of probability functions 81, 46–50
  Schrödinger 206, 210–1, 213, 228–9
  unitary 211, 219–20
expectation value 62–3, 152, 153, 155–6,
  158–60, 193, 226–7
  definition 246, 252